Springer Series on

Wave Phenomena 12

Edited by L. B. Felsen

Springer Series on

Wave Phenomena

J. A. DeSanto

Scalar Wave Theory

Green's Functions and Applications

With 48 Figures

Springer-Verlag
Berlin Heidelberg New York
London Paris Tokyo
Hong Kong Barcelona
Budapest

Professor John A. DeSanto, Ph. D.
Department of Mathematical and Computer Sciences
Colorado School of Mines, Golden, CO 80401, USA

Series Editors:

Professor Leonid M. Brekhovskikh, Academician
Russian Academy of Sciences, P. P. Shirshov Institute of Oceanology,
23 Krasikowa St., 117218 Moscow, Russia

Professor Leopold B. Felsen, Ph.D.
Department of Electrical Engineering, Polytechnic University, Route 110,
Farmingdale, NY 11735, USA

Professor Hermann A. Haus
Department of Electrical Engineering & Computer Sciences, MIT,
Cambridge, MA 02139, USA

Managing Editor: Helmut K. V. Lotsch
Springer-Verlag, Tiergartenstrasse 17,
W-6900 Heidelberg, Fed. Rep. of Germany

ISBN 3-540-55263-4 Springer-Verlag Berlin Heidelberg New York
ISBN 0-387-55263-4 Springer-Verlag New York Berlin Heidelberg

Printed in the United States of America

Typesetting: Camera ready by author

54/3140-5 4 3 2 1 0 – Printed on acid-free paper

This book is dedicated to the memory

of my parents

Esther Keating DeSanto

and

John Carl DeSanto, M.D.

Preface

This book comprises some of the lecture notes I developed for various one- or two-semester courses I taught at the Colorado School of Mines. The main objective of all the courses was to introduce students to the mathematical aspects of wave theory with a focus on the solution of some specific fundamental problems. These fundamental solutions would then serve as a basis for more complex wave propagation and scattering problems.

Although the courses were taught in the mathematics department, the audience was mainly not mathematicians. It consisted of graduate science and engineering majors with a varied background in both mathematics and wave theory in general. I believed it was necessary to start from fundamental principles of both advanced applied mathematics as well as wave theory and to develop them both in some detail. The notes reflect this type of development, and I have kept this detail in the text. I believe it essential in technical careers to see this detailed development at least once.

This volume consists of five chapters. The first two on Scalar Wave Theory (Chapter 1) and Green's Functions (Chapter 2) are mainly mathematical although in Chapter 1 the wave equation is derived from fundamental physical principles. More complicated problems involving spatially and even temporally varying media are briefly introduced. Full solution of these problems involves approximation or numerical methods. These are not included here. In Chapter 2, the development of the free space Green's function for the wave and Helmholtz equation can serve as a precursor to the development of wave theoretic solutions of inhomogeneous problems since the inhomogeneity is often treated as a perturbation from a homogeneous background.

The physical applications are contained in the latter three chapters on Scalar Plane Wave Scattering (Chapter 3), Spherical Waves Scattering from Planar Boundaries (Chapter 4) and a propagation problem in a Two-Layered Liquid Half-Space (Pekeris Waveguide)(Chapter

5). Chapters 3 and 4 are the simplest scattering problems for plane and spherical wave incidence on a planar interface, and are fundamental problems in mathematical physics. The propagation problem in Chapter 5 is also an elementary example in that the media are homogeneous. It elicits a rich mathematical structure found in more complicated inhomogeneous one-dimensional examples, but buried in the latter problems in the complexity and detailed analytic properties of the eigenfunctions. With homogeneity comes a degree of analytical simplification whereby the results can be perceived unmasked by the details.

Again I stress the necessity of the detail in the learning process and trust that this book will serve its intended function.

Golden, CO *John A. DeSanto*
November 1991

Acknowledgments

Parts of these notes have appeared in several courses I have taught at the Colorado School of Mines, most recently as a one-semester course on Boundary Value Problems in Ocean Acoustics and Seismology. I am grateful to the many students in these courses for their comments both in the courses and on these specific notes. I am grateful to Dr. David Stickler for his comments on Chapter 5. For the manuscript preparation, I was very fortunate. Mrs. Barbara McLenon combined both great skill and great enthusiasm in this job. I am very grateful to her for all she has done.

Contents

1. Scalar Wave Theory

We want to discuss the propagation and scattering of sound. This is a small amplitude disturbance in a compressible fluid. To begin then, we first derive the conservation laws applicable to a fluid, and from these the differential equations governing the sound field.

1.1 Fluid Equations

We derive the conservation laws for the mass density ρ and momentum density ρv_i of a fluid. Here v_i is the vector fluid velocity ($i = 1, 2, 3$). The latter is sometimes written as a current $j_i = \rho v_i$. We generally use index notation although in some instances we also quote the standard vector notation. The geometry is illustrated in Fig. 1.1.

Define the following quantities:

$$\begin{aligned} \psi &= \text{some general conserved quantity/unit volume}, \\ \iiint \psi dV &= \text{total conserved quantity}, \\ q &= \text{external source of } \psi\text{/unit time}, \\ \iiint q dV &= \text{total source of } \psi\text{/unit time}. \end{aligned}$$

The flow of ψ through $d\boldsymbol{S}$ per unit time is

$$\psi \boldsymbol{v} \cdot d\boldsymbol{S} = \psi v_i dS_i \tag{1.1.1}$$

with summation over repeated indices from 1 to 3. This has the dimensions:

$$\frac{\text{conserved quantity}}{\text{volume}} \cdot \frac{l}{t} \cdot l^2 = \frac{\text{conserved quantity}}{t}.$$

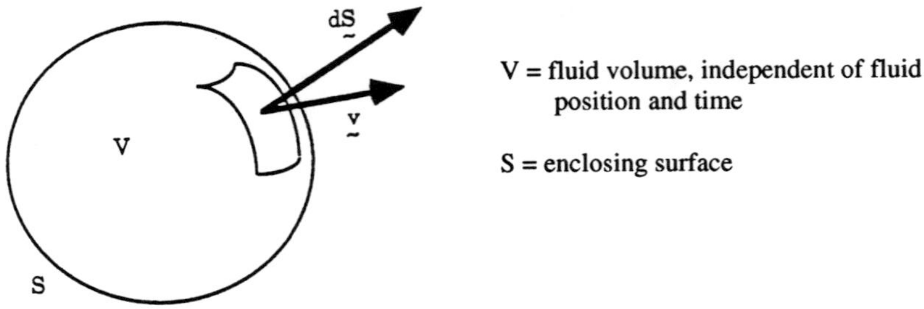

Fig. 1.1. Schematic of fluid volume V surrounded by a surface S with normal dS . The fluid passes through this surface with velocity **v**.

Using Green's theorem the total outflow of ψ can be written as

$$-\int\int \psi \boldsymbol{v} \cdot d\boldsymbol{S} = -\int\int\int \nabla \cdot (\psi \boldsymbol{v}) dV, \tag{1.1.2}$$

or in index notation

$$-\int\int \psi v_i dS_i = -\int\int\int \partial_i(\psi v_i) dV, \tag{1.1.3}$$

where

$$\partial_i = \partial/\partial x_i \quad , \quad i = 1, 2, 3$$

and

$$\partial_t = \partial/\partial t \, .$$

The conservation law for ψ is expressed by the fact that the time rate of change of the conserved quantity is equal to the total source of ψ (inflow) plus the outflow, i.e.

$$\frac{d}{dt}\int\int\int \psi dV = \int\int\int q dV - \int\int\int \partial_i(\psi v_i) dV \, . \tag{1.1.4}$$

In differential form this is given by

$$\partial_t \psi = q - \partial_i(\psi v_i). \tag{1.1.5}$$

For $\psi = \rho$, the mass density, in which case (for our purposes) we choose $q = 0$, is

$$\partial_t \rho + \partial_i(\rho v_i) = 0 \, . \tag{1.1.6}$$

For $\psi = \rho v_j$, the momentum density we get

$$\partial_t(\rho v_j) + \partial_i(\rho v_i v_j) = F_j \ , \tag{1.1.7}$$

where the source of ψ is F_j, the external force densities. The only force of interest here is the force on the fluid due to the local water pressure p given by

$$F_j = -\partial_j p \ . \tag{1.1.8}$$

Substituting (1.1.8) into (1.1.7), expanding the derivative terms and using (1.1.6) yields the equations ($j = 1, 2, 3$)

$$\rho(\partial_t v_j + v_i \partial_i v_j) = -\partial_j p \ , \tag{1.1.9}$$

which are our version of the Euler equations, the fluid version of Newton's second law.

1.2 Sound

Sound is defined as a small amplitude disturbance in a compressible fluid. Compressible means that

$$\partial_j v_j \neq 0 \ . \tag{1.2.1}$$

Equations (1.1.6) and (1.1.9) are four equations in five unknowns, p, ρ and v_j . We add a fifth equation by assuming that the pressure is a function of the density

$$p = p(\rho) \ . \tag{1.2.2}$$

We next expand these five unknowns in a small parameter ε as

$$\rho = \rho^{(0)}(\boldsymbol{x}, t) + \varepsilon \rho^{(1)}(\boldsymbol{x}, t) \ ,$$

$$p = p^{(0)}(\boldsymbol{x}, t) + \varepsilon p^{(1)}(\boldsymbol{x}, t) \ ,$$

and

$$v_j = v_j^{(0)}(\boldsymbol{x}, t) + \varepsilon v_j^{(1)}(\boldsymbol{x}, t), \tag{1.2.3}$$

where the background fluid variables are represented by the superscript (0), and the terms with superscript (1) are the perturbations in the fluid caused by turning on an acoustic source, i.e. some vibrating object whose amplitude and velocity of vibration are small.

Substitute (1.2.3) into (1.1.6), (1.1.9) and (1.2.2) and equate the coefficients of like powers of ε to zero. First, in (1.1.6), the full result is

$$\partial_t \left(\rho^{(0)} + \varepsilon\rho^{(1)}\right) + \partial_i \left\{\left(\rho^0 + \varepsilon\rho^{(1)}\right)\left(v_i^{(0)} + \varepsilon v_i^{(1)}\right)\right\} = 0.$$

The coefficient of ε^0 is given by

$$\varepsilon^0 : \qquad \partial_t\rho^{(0)} + \partial_i \left(\rho^{(0)} v_i^{(0)}\right) = 0.$$

If we assume zero hydrodynamic velocity

$$v_i^{(0)} = 0 \ ,$$

then the equation becomes

$$\partial_t\rho^{(0)} = 0 \ ,$$

which states that $\rho^{(0)}$ is time-independent, viz.,

$$\rho^{(0)} = \rho^{(0)}(\boldsymbol{x}) \ ,$$

so that $\rho^{(0)}$ is at most spatially dependent. To first order in ε the coefficient is

$$\varepsilon : \qquad \partial_t\rho^{(1)} + \rho^{(0)}\partial_i v_i^{(1)} + v_i^{(1)}\partial_i\rho^{(0)} = 0 \ , \tag{1.2.4}$$

where we have again assumed zero hydrodynamic velocity.

Next, substituting in (1.1.9) we get

$$\left[\rho^{(0)} + \varepsilon\rho^{(1)}\right]\left[\varepsilon\partial_t v_j^{(1)} + \mathrm{O}(\varepsilon^2)\right] = -\partial_j \left[p^{(0)} + \varepsilon p^{(1)}\right].$$

The zeroth order coefficient is

$$\varepsilon^0 : \qquad \partial_j p^{(0)} = 0 \ , \tag{1.2.5}$$

which is the case of hydrostatic equilibrium, i.e., to zeroth order there are no external body forces. To first order we have

$$\varepsilon : \qquad \rho^{(0)}\partial_t v_j^{(1)} = -\partial_j p^{(1)} \ . \tag{1.2.6}$$

Finally, expanding about the equilibrium state in (1.2.2) we get

$$\begin{aligned} p^{(0)} + \varepsilon p^{(1)} &= p^{(0)} + \frac{\partial p^{(0)}}{\partial \rho^{(0)}}\left[\rho - \rho^{(0)}\right] \\ &= p^{(0)} + \varepsilon \frac{\partial p^{(0)}}{\partial \rho^{(0)}} \rho^{(1)} \ . \end{aligned}$$

Equating terms of order ε we get

$$p^{(1)} = \frac{\partial p^{(0)}}{\partial \rho^{(0)}} \rho^{(1)} . \tag{1.2.7}$$

Note that if $p^{(0)} = p^{(0)}\left(\rho^{(0)}\right)$, then $p^{(0)} = p^{(0)}(\boldsymbol{x})$, which is time independent since $\rho^{(0)}$ is time independent, and (1.2.7) becomes

$$\partial_j p^{(0)} = \frac{\partial p^{(0)}}{\partial \rho^{(0)}} \partial_j \rho^{(0)}.$$

But by hydrostatic equilibrium $\partial_j p^{(0)} = 0$, hence, $\partial_j \rho^{(0)} = 0$ and we have that both $\rho^{(0)}$ and $p^{(0)}$ are constant. Thus, (1.2.4) becomes

$$\partial_t \rho^{(1)} + \rho^{(0)} \partial_i v_i^{(1)} = 0 . \tag{1.2.8}$$

Substituting (1.2.7) into (1.2.8) we get

$$\partial_t p^{(1)} + \rho^{(0)} \frac{\partial p^{(0)}}{\partial \rho^{(0)}} \partial_i v_i^{(1)} = 0, \tag{1.2.9}$$

which with (1.2.6) form the basic field equations of acoustics, analogous to the Maxwell equations of electromagnetic theory.

1.3 Wave Equations

We first derive a wave equation on the pressure. Take the partial derivative with respect to time of (1.2.9) and subtract from it the gradient (∂_j) of (1.2.6). The result can be written as

$$\nabla^2 p^{(1)} - \frac{1}{c^2} \partial_t^2 p^{(1)} = 0 , \tag{1.3.1}$$

where $\nabla^2 = \partial_j \partial_j$ is the Laplacian (repeated subscript summed from 1 to 3) and where we have defined c as

$$c = \left[\partial p^{(0)} / \partial \rho^{(0)}\right]^{1/2} \tag{1.3.2}$$

which has the dimensions of a (sound) speed, and here is a constant. Equation (1.3.1) is a wave equation on the sound pressure.

To derive a wave equation on the sound velocity, take the divergence of (1.2.9) and subtract from the result the time derivative of (1.2.6) to get

$$\partial_j \partial_i v_i^{(1)} - \frac{1}{c^2} \partial_t^2 v_j^{(1)} = 0 . \tag{1.3.3}$$

The first term is the gradient of a divergence. Recall the vector identity

$$\boldsymbol{\nabla} \times \boldsymbol{\nabla} \times \boldsymbol{v}^{(1)} = \boldsymbol{\nabla} \left[\boldsymbol{\nabla} \cdot \boldsymbol{v}^{(1)} \right] - \nabla^2 \boldsymbol{v}^{(1)} .$$

If we assume that the sound velocity is irrotational

$$\boldsymbol{\nabla} \times \boldsymbol{v}^{(1)} = 0 ,$$

then we can replace this first term by the Laplacian to get the same wave equation for each velocity component

$$\nabla^2 v_j^{(1)} - \frac{1}{c^2} \partial_t^2 v_j^{(1)} = 0 , \tag{1.3.4}$$

which is the same equation as for the pressure. Since the velocity is irrotational, we can write it as the gradient of a scalar (velocity) potential ψ as

$$v_j^{(1)} = -\partial_j \psi , \tag{1.3.5}$$

(the minus sign is a convention). The pressure $p^{(1)}$ can also be defined in terms of ψ as

$$p^{(1)} = \rho^{(0)} \partial_t \psi \tag{1.3.6}$$

(Whatever the sign convention, (1.3.5) and (1.3.6) must be compatible with (1.2.6).) ψ satisfies the same wave equation

$$\nabla^2 \psi - \frac{1}{c^2} \partial_t^2 \psi = 0 . \tag{1.3.7}$$

We generally work with this velocity potential for scalar problems rather than the explicit pressure or velocity. Both are determined from it using (1.3.5) and (1.3.6).

1.4 Spatially Variable Density

We derive a wave equation on the pressure under the assumption that the density is spatially dependent, $\rho^{(0)} = \rho^{(0)}(\boldsymbol{x})$, but independent of time. This changes the sound speed definition (1.3.2) to include this spatial dependence

$$\frac{\partial p^{(0)}}{\partial \rho^{(0)}} = c^2(\boldsymbol{x}) . \tag{1.4.1}$$

In deriving the field equation (1.2.9), we dropped a term (in (1.2.4)) which was the gradient of $\rho^{(0)}$. Consistency demands that we go back to (1.2.4), reinsert the density term and then proceed. We do this in Appendix 1.A and quote the result below. But first we consider what we get if we simply take (1.2.6) and (1.2.9) as given. Differentiate (1.2.9) with respect to time and use (1.4.1) to get

$$\partial_t^2 p^{(1)} + \rho^{(0)} c^2(\boldsymbol{x}) \partial_i \partial_t v_i^{(1)} = 0 ,$$

and insert the time derivative of the velocity from (1.2.6). The result is

$$\rho^{(0)} \partial_j \left\{ \frac{1}{\rho^{(0)}} \partial_j p^{(1)} \right\} - \frac{1}{c^2(\boldsymbol{x})} \partial_t^2 p^{(1)} = 0 . \tag{1.4.2}$$

The consistent derivation done in Appendix A yields an additional term

$$\rho^{(0)} \partial_j \left\{ \frac{1}{\rho^{(0)}} \partial_j p^{(1)} \right\} - \frac{1}{c^2(\boldsymbol{x})} \partial_t^2 p^{(1)} - M(\boldsymbol{x}) p^{(1)} = 0 , \tag{1.4.3}$$

where

$$M(\boldsymbol{x}) = \partial_i \left\{ \frac{1}{\rho^{(0)}} \partial_i \rho^{(0)} \right\} . \tag{1.4.4}$$

The term in brackets in (1.4.4) is the fractional change in density. Its divergence is considered to be very small and for most problems is ignored.

The corresponding equation on the velocity potential can be found by substituting (1.3.6) in (1.4.3) to yield

$$\frac{1}{\rho^{(0)}} \partial_j \left[\rho^{(0)} \partial_j \psi \right] - \frac{1}{c^2(\boldsymbol{x})} \partial_t^2 \psi = 0 , \tag{1.4.5}$$

which contains the density term in the spatial derivative in a different form from its occurrence in (1.4.2) and (1.4.3).

1.5 Helmholtz Equation and Separable Solutions

The equations we derived in sections (1.3) and (1.4) are wave equations and the solutions depend on both space and time. We consider the simplest equation on the velocity potential ψ which is (1.3.7)

$$\nabla^2 \psi - \frac{1}{c^2} \partial_t^2 \psi = 0 ,$$

where c is constant. If we assume a harmonic time dependence and separate the time-dependence out as

$$\psi(\boldsymbol{x}, t) = \psi(\boldsymbol{x}) e^{-i\omega t} \quad , \tag{1.5.1}$$

where ω is circular frequency, then $\psi(\boldsymbol{x})$ satisfies a Helmholtz equation

$$\left(\nabla^2 + k^2\right) \psi(\boldsymbol{x}) = 0 \quad , \tag{1.5.2}$$

where $k = \omega/c$ is the wave number. Implicitly $\psi(\boldsymbol{x})$ is a function of ω, but this is considered a parametric dependence and is suppressed. The full time-dependent solution can be found by a Fourier synthesis over frequency of the time-independent frequency-parametric solution $\psi(\boldsymbol{x})$. Solutions to (1.5.2) can be found using several methods. We briefly review the solutions found using the method of separation of variables for Cartesian, cylindrical and spherical coordinates.

For *Cartesian coordinates*, (1.5.2) becomes

$$\partial_x^2 \psi + \partial_y^2 \psi + \partial_z^2 \psi + k^2 \psi = 0 \quad . \tag{1.5.3}$$

A separable solution is found in the form

$$\psi(\boldsymbol{x}) = \mathrm{X}(x)\mathrm{Y}(y)\mathrm{Z}(z) \quad . \tag{1.5.4}$$

This is substituted into (1.5.3) and the result divided by ψ to yield

$$\frac{1}{\mathrm{X}}\mathrm{X}'' + \frac{1}{\mathrm{Y}}\mathrm{Y}'' + \frac{1}{\mathrm{Z}}\mathrm{Z}'' + k^2 = 0 \quad .$$

If we set

$$\mathrm{X}'' = -\alpha^2\, \mathrm{X}, \mathrm{Y}'' = -\beta^2 \mathrm{Y},\ \mathrm{Z}'' = -\gamma^2\, \mathrm{Z} \quad ,$$

then the sets of solutions are (a_j and b_j are constants)

$$\mathrm{X}(x) = \begin{cases} \{\sin \alpha x, \cos \alpha x\} \text{ or } \{\exp(\pm i\alpha x)\} \ , & \alpha \neq 0 \\ a_1 x + b_1 \ , & \alpha = 0 \end{cases}$$

$$\mathrm{Y}(y) = \begin{cases} \{\sin \beta y, \cos \beta y\} \text{ or } \{\exp(\pm i\beta y)\} \ , & \beta \neq 0 \\ a_2 y + b_2 \ , & \beta = 0 \end{cases}$$

and

$$\mathrm{Z}(z) = \begin{cases} \{\sin \gamma z, \cos \gamma z\} \text{ or } \{\exp(\pm i\gamma z)\} \ , & \gamma \neq 0 \\ a_3 z + b_3 \ , & \gamma = 0 \end{cases}$$

where we have the constraint

$$k^2 = \alpha^2 + \beta^2 + \gamma^2 \ .$$

For example, for a box, $0 \le x \le a$, $0 \le y \le b$, $0 \le z \le d$, with ψ vanishing on all sides (trivial case) the eigenfunction solutions are

$$\psi_{nml} = \sin(n\pi x/a)\sin(m\pi y/b)\sin(l\pi z/d) \ ,$$

with $\alpha = n\pi/a$, $\beta = m\pi/b$ and $\gamma = l\pi/d$ where $n, m, l = 1, 2 \cdots$. The eigenvalue spectrum is discrete and can be characterized by the values

$$k_{nml} = \left[(n\pi/a)^2 + (m\pi/b)^2 + (l\pi/d)^2\right]^{1/2} \ .$$

The full solution is found from a superposition of these functions. We discuss this further below. If any or all of the sides of the box approach infinity, the eigenvalue spectrum becomes continuous, since the positions of the eigenvalues bunch up. The discrete points come together to form a continuum. Simple examples of these latter eigenfunctions are plane waves. In one dimension

$$e^{i(\alpha x - \omega t)}$$

is a wave that moves to the right (positive x-direction) and

$$e^{-i(\alpha x + \omega t)}$$

moves to the left. In three dimensions we have

$$e^{i(\boldsymbol{k}\cdot\boldsymbol{x} - \omega t)} \ ,$$

a plane wave progressing in the **k**-direction which can be written as

$$\boldsymbol{k} = k(\alpha, \beta, \gamma) = k(\sin\theta\cos\phi,\ \sin\theta\sin\phi,\ \cos\theta) \ ,$$

where θ and ϕ are angles in spherical coordinates. The surfaces of constant phase are planar. The full solution of the problem is a superposition over those eigenfunctions which satisfies the appropriate boundary value problem.

In *cylindrical coordinates* (1.5.2) becomes

$$\frac{1}{r}\frac{\partial}{\partial r}\left(r\frac{\partial\psi}{\partial r}\right) + \frac{1}{r^2}\frac{\partial^2\psi}{\partial\phi^2} + \frac{\partial^2\psi}{\partial z^2} + k^2\psi = 0 \ . \qquad (1.5.5)$$

If we separate out the z-dependence first as

$$\psi(r,\phi,z) = F(r,\phi)f(z) \ , \tag{1.5.6}$$

then substitute (1.5.6) in (1.5.5) and divide the result by (1.5.6) we get

$$\frac{1}{rF}\frac{\partial}{\partial r}\left(r\frac{\partial F}{\partial r}\right) + \frac{1}{r^2F}\frac{\partial^2 F}{\partial \phi^2} + \frac{1}{f}\frac{d^2 f}{dz^2} + k^2 = 0 \ . \tag{1.5.7}$$

We choose the separability

$$f'' + p^2 f = 0 \ , \tag{1.5.8}$$

with the class of solutions

$$f : \left\{ \begin{array}{lcl} \exp(\pm ipz)\ \mathrm{or}(\cos pz, \sin pz) & : & p \neq 0 \\ 1,\ z & : & p = 0 \end{array} \right\} \ . \tag{1.5.9}$$

Further separability

$$F(r,\phi) = R(r)\Phi(\phi) \ , \tag{1.5.10}$$

and the choice

$$\Phi'' + m^2\Phi = 0 \ , \tag{1.5.11}$$

with solutions

$$\Phi : \left\{ \begin{array}{lcl} \exp(\pm im\phi)\ \mathrm{or}\ (\cos m\phi, \sin m\phi) & : & m \neq 0 \\ 1, \phi & : & m = 0 \end{array} \right\} \ , \tag{1.5.12}$$

yields the radial equation

$$R'' + \frac{1}{r}R' + \left[k^2 - l^2 - \frac{m^2}{r^2}\right] R = 0 \ , \tag{1.5.13}$$

which has solutions

$$R(r) = \left\{ \begin{array}{lll} Z_m([k^2 - l^2]^{1/2} r) & : & k \neq l \quad (\textit{Bessel functions}) \\ r^{\pm m} & : & k = l, \quad m \neq 0 \\ 1, \log r & : & k = 1, \quad m = 0 \ . \end{array} \right. \tag{1.5.14}$$

The term m is usually chosen to be an integer so that the solutions repeat as ϕ goes from 0 to 2π . The separation as described above is a $k_z - $*separation* in that p in (1.5.8) is the z-component of wavenumber. An alternate $k_r -$ *separation* can be found if we let

$$\beta = [k^2 - l^2]^{1/2} \ , \tag{1.5.15}$$

which is the radial component of wavenumber. In this case the combined radial and depth solutions for $\beta \neq 0$ are

$$Z_m(\beta r) \exp\left\{\pm i[k^2 - \beta^2]^{1/2} z\right\} \ . \tag{1.5.16}$$

For m an integer the full radial solutions can be written as either (C_j constants)

$$R(r) = \begin{cases} C_1 J_m(\beta r) + C_2 N_m(\beta r) \ , \\ \\ C_1 H_m^{(1)}(\beta r) + C_2 H_m^{(2)}(\beta r) \ , \end{cases} \tag{1.5.17}$$

where J_m is the usual Bessel function, N_m is the Neumann function and

$$H_m^{(1,2)}(\beta r) = J_m(\beta r) \pm i N_m(\beta r) \ , \tag{1.5.18}$$

are the Hankel functions of first and second kinds respectively. We have the following asymptotic properties which will be useful

$$r \to 0 : \begin{cases} J_0(\beta r) \sim 1 \ , \\ \\ J_m(\beta r) \sim (\beta r)^m \qquad m \neq 0 \ , \\ \\ N_0(\beta r) \sim \log r \ , \\ \\ N_m(\beta r) \sim (\beta r)^{-m} \quad m \neq 0 \ , \end{cases} \tag{1.5.19}$$

and

$$r \to \infty : \begin{cases} J_m(\beta r) \sim \left(\frac{2}{\pi \beta r}\right)^{1/2} \cos\left(\beta r - \frac{m\pi}{2} - \frac{\pi}{4}\right) \ , \\ \\ N_m(\beta r) \sim \left(\frac{2}{\pi \beta r}\right)^{1/2} \sin\left(\beta r - \frac{m\pi}{2} - \frac{\pi}{4}\right) \ , \\ \\ H_m^{(1,2)}(\beta r) \sim \left(\frac{2}{\pi \beta r}\right)^{1/2} \exp\left\{\pm i\left[\beta r - \frac{m\pi}{2} - \frac{\pi}{4}\right]\right\} \ . \end{cases} \tag{1.5.20}$$

As $r \to \infty$, J_m and N_m behave like standing waves, whereas the Hankel functions behave like outgoing ($\exp(i\beta r)$) and incoming ($\exp(-i\beta r)$) radial waves respectively.

In *spherical coordinates* (r, θ, ϕ) (1.5.2) becomes

$$\frac{1}{r^2}\frac{\partial}{\partial r}\left(r^2 \frac{\partial \psi}{\partial r}\right) + \frac{1}{r^2 \sin\theta}\frac{\partial}{\partial \theta}\left(\sin\theta \frac{\partial \psi}{\partial \theta}\right) + \frac{1}{r^2 \sin^2\theta}\frac{\partial^2 \psi}{\partial \phi^2} + k^2 \psi = 0 \ , \tag{1.5.21}$$

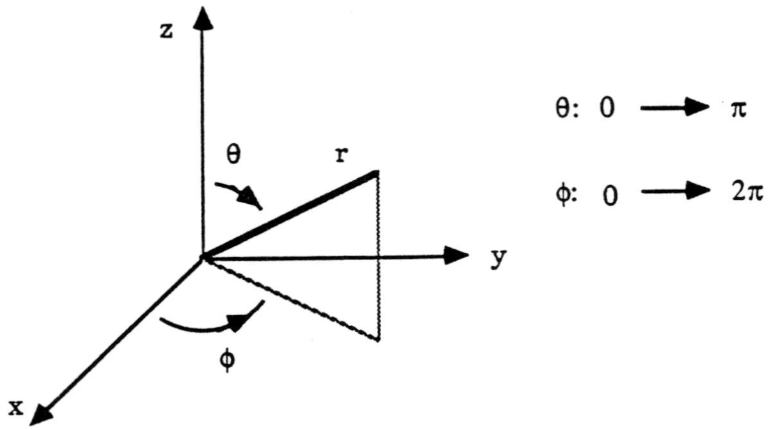

Fig. 1.2. Spherical geometry.

where the geometry is defined in Fig. 1.2. Separating the radial part

$$\psi(r,\theta,\phi) = R(r)Y(\theta,\phi) \ , \tag{1.5.22}$$

yields

$$\frac{1}{R}\frac{d}{dr}\left(r^2\frac{dR}{dr}\right) + k^2r^2 + \frac{1}{Y\sin\theta}\frac{\partial}{\partial\theta}\left(\sin\theta\frac{\partial Y}{\partial\theta}\right) + \frac{1}{Y\sin^2\theta}\frac{\partial^2 Y}{\partial\phi^2} = 0 \ . \tag{1.5.23}$$

Choose the separation constant as the product $l(l+1)$ where l is (usually) a positive integer or zero. We choose it so here. This separation arises essentially because we require the solutions in θ to be finite at both $\theta = 0$ and $\theta = \pi$. Series solutions of the θ-equation quoted below yield finite solutions if they are polynomials, in which case the general separation constant has the form $l(l+1)$. Then the radial equation is

$$\frac{d}{dr}\left(r^2\frac{dR}{dr}\right) + \left(k^2r^2 - l(l+1)\right)R = 0 \ , \tag{1.5.24}$$

which has spherical Bessel function solutions. We define the spherical Bessel, Neumann, and Hankel functions as ($x = kr$)

$$\begin{aligned} j_l(x) &= \left(\frac{\pi}{2x}\right)^{1/2} J_{l+1/2}(x) \ , \\ n_l(x) &= \left(\frac{\pi}{2x}\right)^{1/2} N_{l+1/2}(x) \ , \end{aligned}$$

and

$$h_l^{(1,2)}(x) = j_l(x) \pm i\, n_l(x) \quad , \tag{1.5.25}$$

which have asymptotic properties

$$x \to 0 : \begin{cases} j_l(x) \sim x^l \quad , \\ \\ n_l(x) \sim x^{-l-1} \quad , \end{cases} \tag{1.5.26}$$

and

$$x \to \infty : \begin{cases} j_l(x) \sim \frac{1}{x} \cos\left[x - (l+1)\frac{\pi}{2}\right] \quad , \\ \\ n_l(x) \sim \frac{1}{x} \sin\left[x - (l+1)\frac{\pi}{2}\right] \quad , \\ \\ h_l^{(1)}(x) \sim \frac{e^{ix}}{x} \begin{cases} -i(-1)^{l/2} & l \textit{ even} \\ (-1)^{(l+1)/2} & l \textit{ odd} \quad , \end{cases} \\ \\ h_l^{(2)}(x) \sim \frac{e^{-ix}}{x} \begin{cases} i(-1)^{l/2} & l \textit{ even} \\ (-1)^{(l+1)/2} & l \textit{ odd} \quad . \end{cases} \end{cases} \tag{1.5.27}$$

Again, as $x \to \infty$, $h_l^{(1)}$ and $h_l^{(2)}$ behave as outgoing and incoming waves respectively.

Separation of the angular variability

$$Y(\theta,\phi) = F(\theta)\Phi(\phi) \quad , \tag{1.5.28}$$

the choice of the differential equation for Φ as (1.5.11), and solutions (1.5.12) yields an equation on F

$$\sin\theta \frac{d}{d\theta}\left(\sin\theta \frac{dF}{d\theta}\right) + \left[l(l+1)\sin^2\theta - m^2\right] F = 0 \quad . \tag{1.5.29}$$

If we let

$$x = \cos\theta$$

$$\frac{d}{d\theta} = \frac{dx}{d\theta}\frac{d}{dx} = -\sin\theta \frac{d}{dx}$$

$$\sin\theta \frac{d}{d\theta} = -(1-x^2)\frac{d}{dx} \quad ,$$

then the equation becomes

$$\frac{d}{dx}\left[(1-x^2)\frac{dF}{dx}\right]+\left[l(l+1)-\frac{m^2}{1-x^2}\right]F=0 \ . \tag{1.5.30}$$

This is the Legendre differential equation which for $m = 0$ has Legendre polynomial solutions $P_l(x)$ and for $m \neq 0$ solutions $P_l^m(x)$, the associated Legendre functions. We can further define the spherical (surface) harmonics

$$Y_{ml}(\theta,\phi)=\exp(im\phi)P_l^m(\cos\theta) \ , \tag{1.5.31}$$

combining the angular variability. These satisfy the orthogonality relation

$$\int_0^{2\pi} d\phi \int_0^{\pi} d\theta \sin\theta \, Y_{ml}(\theta,\phi)Y^*_{m'l'}(\theta,\phi) = \delta_{mm'}\,\delta_{ll'}\,\frac{4\pi}{2l+1}\,\frac{(l+|m|)!}{(l-|m|)!} \ , \tag{1.5.32}$$

where the $*$ represents complex conjugation. Here $|m| \leq l$.

1.6 One-Dimensional Problems and Sturm-Liouville Theory

The separation of variables technique, where applicable, enables us to compose the solution of multidimensional problems from those of one-dimensional problems. In this section we discuss the properties of a general class of one-dimensional problems which are referred to as Sturm-Liouville problems. We begin with a second order ordinary differential equation of general form

$$\frac{d}{dz}\left[p(z)\frac{d\psi}{dz}\right]+\left[q(z)+\lambda r(z)\right]\psi=0 \ ,$$

where p, q and r are arbitrary functions whose properties are specified further below. The parameter λ will be adjusted to satisfy the boundary conditions on ψ, also specified below. The permissible values of λ form what is termed the *eigenvalue spectrum*, which can consist of discrete values λ_n (where n is an integer) and/or a continuous set of eigenvalues. To each eigenvalue there is a corresponding eigenfunction ψ_n. We discuss the properties of the eigenfunctions and the spectrum below. Some properties we prove, while others we merely state.

(A) The Eigenfunctions Are Orthogonal To prove this for $z\varepsilon[a,b]$ we write the corresponding differential equations for the nth and the mth eigenfunctions which are (the prime means differentiation in z)

$$(p\psi_n')' + (q + \lambda_n r)\psi_n = 0 \ , \tag{1.6.1}$$

$$(p\psi_m')' + (q + \lambda_m r)\psi_m = 0 \ . \tag{1.6.2}$$

Multiply (1.6.1) by ψ_m and (1.6.2) by ψ_n and subtract the resulting equations. The result can be written as

$$\frac{d}{dz}\{p\psi_m\psi_n' - p\psi_n\psi_m'\} + (\lambda_n - \lambda_m)r(z)\psi_m\psi_n = 0 \ . \tag{1.6.3}$$

Integrate this equation from a to b to get

$$\{p\psi_m\psi_n' - p\psi_n\psi_m'\}_a^b + (\lambda_n - \lambda_m)\int_a^b r(z)\psi_m\psi_n dz = 0 \ . \tag{1.6.4}$$

The *Sturm-Liouville (S-L) boundary conditions* consist in setting the bracket to zero, i.e.,

$$\begin{aligned}\{p\psi_m\psi_n' - p\psi_n\psi_m'\}_a^b &= p(b)\left[\psi_m\psi_n' - \psi_n\psi_m'\right]_b \\ &\quad -p(a)\left[\psi_m\psi_n' - \psi_n\psi_m'\right]_a = 0 \ .\end{aligned} \tag{1.6.5}$$

Then if $\lambda_n \neq \lambda_m$, the eigenfunctions are *orthogonal with respect to the weight function* $r(z)$, i.e.,

$$\int_a^b r(z)\psi_m(z)\psi_n(z)dz = 0. \tag{1.6.6}$$

If there are m linearly independent solutions corresponding to the same eigenvalue, i.e. there is what is termed an m-fold degeneracy, then the orthogonalization can be accomplished using the Gram-Schmidt orthogonalization process discussed elsewhere.

The types of boundary conditions for which (1.6.5) is satisfied are the following:

1. $\psi = 0$ at $z = a, b$ (Dirichlet conditions)

2. $\psi' = 0$ at $z = a, b$ (Neumann conditions)

3. $\psi + \alpha\psi' = 0(\alpha\ real)$ at $z = a, b$ (Mixed conditions or Robin conditions)

   An example which does *not* fit this condition is the impedance boundary condition where the impedance is defined as the ratio of pressure to normal velocity (for a flat boundary)

$$Z = p^{(1)}/v_n^{(1)} \ .$$

   For a flat boundary in z, this can be written using (1.3.5) and (1.3.6) as

$$Z = \rho^{(0)}\partial\psi/\partial t\,\{\partial\psi/\partial z\}^{-1} \ ,$$

   where the normal is in the negative z-direction (see Sect. 3.3) and which for time harmonic problems is

$$Z = -i\omega\rho^{(0)}\psi/\psi' \ .$$

   For a real impedance the ratio ψ/ψ' is imaginary.

4. Any of the above three boundary conditions at either boundary.
5. Periodic coefficients $p(a) = p(b)$, and periodic boundary conditions

$$\psi(a) = \psi(b)\ ; \qquad \psi'(a) = \psi'(b) \ .$$

6. Periodic coefficients $p(a) = p(b)$ and anti-periodic boundary conditions

$$\psi(a) = -\psi(b)\ ; \qquad \psi'(a) = -\psi'(b) \ .$$

7. Periodic coefficients $p(a) = p(b)$ and boundary conditions

$$\psi'(a) = -\psi(b)\ ; \qquad \psi'(b) = \psi(a) \ .$$

8. Periodic coefficients $p(a) = p(b)$ and boundary conditions

$$\psi'(a) = \psi(b)\ ; \qquad \psi'(b) = -\psi(a) \ .$$

9. An exceptional case at one boundary. For example, let $p(a) = 0$. Then we need only require that $\psi(a)$ and $\psi'(a)$ be finite. These latter are weaker boundary conditions and can lead to a continuous spectrum.

(B) The Eigenvalues Are Real Assume that p, q, r are real and $r > 0$. Then the equation for the nth eigenfunction is (1.6.1),

$$(p\psi_n')' + (q + \lambda_n r)\psi_n = 0 \ ,$$

and the equation for its complex conjugate is

$$(p\psi_n^{*'})' + (q + \lambda_n^* r)\psi_n^* = 0 \ . \tag{1.6.7}$$

Multiply (1.6.1) by ψ_n^* and (1.6.7) by ψ_n. Subtracting the resulting equations and integrating the result from a to b yields with the Sturm-Liouville boundary conditions The integral is always positive since $r > 0$ and thus

$$\lambda_n = \lambda_n^* \ , \tag{1.6.8}$$

which proves the assertion.

(C) Proper S-L Problem A proper S-L problem is defined as having the following conditions:

1. p', q, and r are real continuous functions on the finite interval $[a, b]$.
2. $p > 0$, $r > 0$ and $q \leq 0$ in $[a, b]$ with the S-L boundary conditions imposed at a, b. We exclude the boundary condition number 9.

This leads to the following.

Theorem 1.1 *A proper S-L problem has a discrete spectrum of eigenvalues λ_0, λ_1, $\cdots$ having minimum eigenvalue λ_0, and no upper bound or limit point.*

Essentially a proper problem refers to real coefficients and a finite region. Complex coefficients or an infinite boundary form an improper problem which can lead to a continuous spectrum. For the normalization of the eigenfunctions for these problems we choose

$$\int_a^b r(z)\psi_m(z)\psi_n^*(z)dz = \delta_{mn} \ . \tag{1.6.9}$$

The normalization is also true for more general cases where the eigenvalues are complex.

(D) The Eigenvalues Are Positive This result can be stated using the following.

Theorem 1.2

Given a S-L system, i.e. the differential equation

$$\frac{d}{dz}\left(p\frac{d\psi}{dz}\right) + (q + \lambda r)\psi = 0 \ , \tag{1.6.10}$$

and the boundary conditions of the form

$$\alpha'\psi'(a) - \alpha\psi(a) = 0 \ ,$$

$$\beta'\psi'(b) - \beta\psi(b) = 0 \ , \tag{1.6.11}$$

where α, α', β, and β' are all real. We assume p, q, r are real continuous functions in $[a, b]$, p' continuous in $[a, b]$, $p > 0$, $r > 0$ and $q \leq 0$ in $[a, b]$, and p, q, r, α, α', β, and β' are independent of λ. Then there exists a set of real numbers $\lambda_0, \lambda_1, \cdots$ which have no limit except $\lambda = \infty$, and a corresponding set of eigenfunctions $\psi_0, \psi_1, \cdots$ such that ψ_m has exactly m zeroes in (a, b). If in addition $\alpha\,\alpha' \geq 0$ and $\beta\,\beta' \geq 0$, then $\lambda_j > 0$ for all j.

(E) Expansion Theorem Using these eigenfunctions we can define an *expansion theorem* as follows: given a function $f(z)$, expand it in a series of eigenfunctions

$$f(z) = \sum_n a_n\psi_n(z) \ . \tag{1.6.12}$$

To find the expansion coefficients a_n we multiply both sides by the integral operator

$$\int_a^b r(z)\psi_m^*(z)dz \ ,$$

and use the orthogonality of the eigenfunctions given by (1.6.10). Term by term integration is justified if the expansion is absolutely and uniformly convergent. Then the coefficients a_n are given by

$$a_n = \int_a^b r(z)\psi_n^*(z)f(z)dz \ . \tag{1.6.13}$$

Specifically, we have the following.

Theorem 1.3

Given $f(z)$ a bounded, piecewise continuous function with piecewise continuous first and second derivatives in $[a, b]$. The set $\{\psi_n\}$ are the eigenfunctions of a proper S-L problem. Then:

$$\sum_{n=0}^{\infty} a_n \psi_n(z),$$

with a_n given by (1.6.14) converges uniformly and absolutely to:

1. *$f(z)$ where $z \varepsilon (a, b)$ and z is not a point of discontinuity of f.*
2. *$\frac{1}{2}[f(z_+) + f(z_-)]$ at a point of discontinuity of f. Here $f(z_+)$ and $f(z_-)$ are the right and left limits of f respectively.*

(F) Wronskian Given two differentiable functions ψ_1 and ψ_2 on some open interval (a, b), the following determinant,

$$W = W(\psi_1, \psi_2) = \begin{vmatrix} \psi_1 & \psi_2 \\ \psi_1' & \psi_2' \end{vmatrix} = \psi_1 \psi_2' - \psi_2 \psi_1' \ , \tag{1.6.14}$$

is called the Wronskian of the two functions. It is important because of the following result. If ψ_1, ψ_2 are solutions of a S-L problem and if there exists at least one point in (a, b) where $W \neq 0$, then the two solutions are *linearly independent* and the general solution of the equation, $\psi(z)$, can be written as a linear combination of them, viz.,

$$\psi(z) = C_1 \psi_1(z) + C_2 \psi_2(z) \ , \tag{1.6.15}$$

where C_1 and C_2 are constants.

If ψ satisfies the boundary conditions (1.6.12), then substituting (1.6.16) in (1.6.12) yields

$$\begin{aligned} C_1\Big[\alpha' \psi_1'(a) - \alpha \psi_1(a)\Big] + C_2\Big[\alpha' \psi_2'(a) - \alpha \psi_2(a)\Big] &= 0 \ , \\ C_1\Big[\beta' \psi_1'(b) - \beta \psi_1(b)\Big] + C_2\Big[\beta' \psi_2'(b) - \beta \psi_2(b)\Big] &= 0 \ , \end{aligned} \tag{1.6.16}$$

which is a set of linear homogeneous equations for the constants C_1 and C_2. They have a non-trivial solution provided the determinant of coefficients vanishes,

$$\begin{vmatrix} \alpha' \psi_1'(a) - \alpha \psi_1(a) & \alpha' \psi_2'(a) - \alpha \psi_2(a) \\ \beta' \psi_1'(b) - \beta \psi_1(b) & \beta' \psi_2'(b) - \beta \psi_2(b) \end{vmatrix} = 0 \ . \tag{1.6.17}$$

This is the *characteristic equation*, and its solution yields the *eigenvalues* λ_n of the problem.

(G) Mean Convergence Another, more practical, convergence criterion for a linear superposition of eigenfunctions is the following: Assume ψ_n are the eigenfunctions of a proper S-L problem. Given a linear combination of the form

$$\sum_{m=0}^{n} C_m \psi_m(z) \ , \tag{1.6.18}$$

form the quantity

$$I_n = \int_a^b \left[f(z) - \sum_{m=0}^{n} C_m \psi_m(z) \right]^2 dz \ . \tag{1.6.19}$$

Then the linear combination *converges in the mean* to $f(z)$ iff

$$\lim_{n\to\infty} I_n = 0 \ , \tag{1.6.20}$$

i.e., for every $\varepsilon > 0$, there exists an $n > n_0$ such that $I_n < \varepsilon$. The limit is written

$$f(z) = \underset{n\to\infty}{\text{l.i.m.}} \sum_{m=0}^{n} C_m \psi_m(z) \ , \tag{1.6.21}$$

where l.i.m. means the *limit in the mean.* The set $\{\psi_n\}$ is said to be *complete* iff $C_n = a_n$ where a_n is given by (1.6.14), or iff the *Parseval identity*

$$\sum_{n=0}^{\infty} C_n^2 = \int_a^b f^2(z) dz \ , \tag{1.6.22}$$

is true. This is a weaker condition than absolute and uniform convergence, because here the convergence is not pointwise.

(H) Completeness Relation If we are given the expansion

$$f(z) = \sum_n a_n \psi_n(z) \ , \tag{1.6.23}$$

and we substitute for a_n from (1.6.14) we have

$$f(z) = \int_a^b dz' \left\{ \sum_n r(z') \psi_n^*(z') \psi_n(z) \right\} f(z') \ . \tag{1.6.24}$$

The term in brackets must be such as to reproduce the function f. This is true if

$$r(z') \sum_n \psi_n^*(z') \psi_n(z) = \delta(z' - z) \ , \tag{1.6.25}$$

which is called a completeness relation for the S-L problem.

(I) Continuous Spectrum Up to now we have considered problems with discrete eigenvalues which is true for a proper S-L problem on a finite interval. In the box problem in Sec.(1.5) we noted that if any dimension of the box should become infinite the eigenvalues will bunch together to form a continuum or a continuous spectrum. The same is true if the coefficients of our differential equation in the S-L problem are singular or vanish at the endpoints of the boundary value problem. In any of these cases we get what is referred to as a singular S-L problem. Briefly, we also then have an additional continuous set of eigenfunctions $\psi(z,\lambda)$, which are normalized as

$$\int r(z)\psi^*(z,\lambda')\psi(z,\lambda)dz = \delta(\lambda'-\lambda) \ , \tag{1.6.26}$$

and any expansion theorem must include them, e.g.

$$f(z) = \sum_m a_m\psi_m(z) + \int_C b(\lambda)\psi(z,\lambda)d\lambda \ . \tag{1.6.27}$$

The $\{\psi_m\}$ are referred to as *proper eigenfunctions* and the functions $\psi(z,\lambda)$ as *improper eigenfunctions.* The integral in (1.6.27) is over that continuous region where they exist. We consider some examples in Chap. 5.

1.7 Calculation of Eigenvalues from Eigenfunctions

We showed that, for a proper S-L boundary value problem, the eigenvalues were real. We showed this by cross multiplying (1.6.1) and (1.6.7) and subtracting the resulting equations. To find the real part of λ_n we add the resulting equations to get

$$(p\psi_n')'\psi_n^* + (p\psi_n^{*\prime})'\psi_n + 2q|\psi_n|^2 + (\lambda_n+\lambda_n^*)r|\psi_n|^2 = 0 \ . \tag{1.7.1}$$

Rewrite the first two terms as

$$\begin{aligned}(p\psi_n')'\psi_n^* + (p\psi_n^{*\prime})'\psi_n &= p'(\psi_n'\psi_n^* + \psi_n^{*\prime}\psi_n) + p(\psi_n''\psi_n^* + \psi_n^{*\prime\prime}\psi_n) \\ &= p'\frac{d}{dz}|\psi_n|^2 + p\left(\frac{d^2}{dz^2}|\psi_n|^2 - 2|\psi_n'|^2\right) \\ &= \frac{d}{dz}\left[p\frac{d}{dz}|\psi_n|^2\right] - 2p|\psi_n'|^2 \ . \end{aligned} \tag{1.7.2}$$

Integrate the resulting equation and use the fact that $\lambda_n + \lambda_n^* = 2\mathrm{Re}\lambda_n$. The result can be written as

$$\mathrm{Re}\lambda_n = N_n / D_n \ , \tag{1.7.3}$$

where

$$\begin{aligned} N_n &= \int_a^b p(z)|\psi_n'(z)|^2 dz - \int_a^b q(z)|\psi_n(z)|^2 dz \\ &\qquad - \frac{1}{2}\left[p\frac{d}{dz}|\psi_n|^2\right]_a^b , \end{aligned} \tag{1.7.4}$$

and

$$D_n = \int_a^b r(z)|\psi_n(z)|^2 dz = 1 \ .$$

The latter is just our normalization integral (1.6.10). In (1.7.4) the integrated term on the *rhs* vanishes if we assume the S-L boundary conditions. If, in addition, we assume that $q(z) \leq 0$, we see that $\mathrm{Re}\lambda_n > 0$ from (1.7.3).

The fact that the eigenvalues were real depended on the fact that p, q and r were real, as well as the fact that the boundary conditions were real. For example, in the mixed boundary condition, α was required to be real. If any of the above conditions are violated, the eigenvalues are complex. For example, if p, r and the boundary conditions are real, but if q is complex, (1.6.7) becomes

$$(p\psi_n^{*'})' + (q^* + \lambda_n^* r)\psi_n^* = 0 \ . \tag{1.7.5}$$

Multiplying (1.6.1) by ψ_n^* and subtracting from the result ψ_n times (1.7.5) we get

$$\frac{d}{dz}(p\psi_n^*\psi_n' - p\psi_n^{*'}\psi_n) + (q - q^*)|\psi_n|^2 + (\lambda_n - \lambda_n^*)r|\psi_n|^2 = 0 \ . \tag{1.7.6}$$

Integration from a to b in z and application of real boundary conditions eliminates the first term and yields

$$\mathrm{Im}\,\lambda_n = -\int_a^b \mathrm{Im}\, q(z)|\psi_n(z)|^2 dz \ , \tag{1.7.7}$$

where we have again used the normalization integral (1.6.10).

The real part of the eigenvalues can be found by multiplying (1.6.1) by ψ_n^* and adding to this (1.7.5) times ψ_n to yield

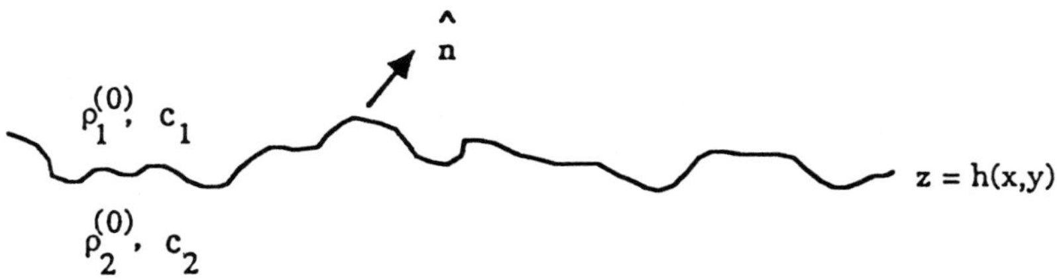

Fig. 1.3. Schematic of a rough interface separating media of different density (ρ_j) and wave speed (c_j).

$$(p\psi_n')'\psi_n^* + (p\psi_n^{*'})'\psi_n + (q+q^*)|\psi_n|^2 + (\lambda_n + \lambda_n^*)r|\psi_n|^2 = 0 \ . \quad (1.7.8)$$

Rewriting the p-terms using (1.7.2) and integrating the result from a to b yields

$$\mathrm{Re}\lambda_n = \int_a^b p(z)|\psi_n'(z)|^2 dz - \int_a^b \mathrm{Re}\, q(z)|\psi_n(z)|^2 dz - \frac{1}{2}\left[p\frac{d}{dz}|\psi_n|^2\right]\Big|_a^b \ , \quad (1.7.9)$$

which is the same as (1.7.4) with q replaced by Req. Again the integrated term vanishes if we use the S-L boundary conditions. Note that (1.7.7) and (1.7.9) are implicit equations for the eigenvalues since the eigenfunctions themselves depend on λ_n.

1.8 Interface Boundary Conditions

At an interface $z = h(x,y)$ separating media of different density ($\rho_j^{(0)}, j = 1,2$) and wave speed (c_j) (regions 1 and 2) where $\hat{n}$ is the normal to the interface, we have the physical boundary conditions of continuity of pressure $p^{(1)}$ and normal component of velocity $\boldsymbol{v}^{(1)}$, i.e. at $z = h$ (see Fig. 1.3)

$$p_1^{(1)} = p_2^{(1)}$$

$$\boldsymbol{v}_1^{(1)} \cdot \hat{n} = \boldsymbol{v}_2^{(1)} \cdot \hat{n} \ . \quad (1.8.1)$$

Expressing the pressure and velocity in terms of the velocity potential ψ using (1.3.5) and (1.3.6), and assuming the problem is time harmonic, these interface conditions at $z = h$ can be written as

$$\rho_1\psi_1 = \rho_2\psi_2 \ , \tag{1.8.2}$$

and

$$\nabla\psi_1 \cdot \hat{n} = \nabla\psi_2 \cdot \hat{n} \ . \tag{1.8.3}$$

For a flat interface the latter condition is just

$$\frac{\partial\psi_1}{\partial z} = \frac{\partial\psi_2}{\partial z} \ . \tag{1.8.4}$$

We use these latter conditions extensively in future chapters. (Note specifically that these conditions are written on the velocity potential. It is also possible to write slightly modified conditions on the pressure or impedance.)

1.9 Energy and Momentum

To compute relations involving the energy of the scalar wave we multiply (1.2.9) by $p^{(1)}$ and (1.2.6) by $v_j^{(1)}$ and add the resulting equations. The result can be written as

$$\partial_j S_j + \partial_t W = 0 \ , \tag{1.9.1}$$

where

$$S_j = p^{(1)} v_j^{(1)} \ , \tag{1.9.2}$$

is the energy flux (power/area or energy/area sec) or Poynting vector and

$$W = \frac{1}{2}(\rho^{(0)}c^2)^{-1}(p^{(1)})^2 + \frac{1}{2}\rho^{(0)}(v^{(1)})^2 \ , \tag{1.9.3}$$

is the energy density of the wave (*not* of the fluid). The first term in (1.9.3) is the potential energy density which can be rewritten as

$$\frac{1}{2}\left[\rho^{(0)}c^2\right]^{-1}\left[p^{(1)}\right]^2 = \frac{K}{2}\left[p^{(1)}\right]^2 \tag{1.9.4}$$

where K is the adiabatic compressibility defined by

$$K = -\frac{1}{V}\left(\frac{\partial V}{\partial p^{(0)}}\right)_s \ , \tag{1.9.5}$$

where the change in volume V with respect to pressure is taken at constant entropy s. Since the volume is inversely proportional to $\rho^{(0)}$ we can also write this result as

$$K = \left[\rho^{(0)}\right]^{-1}\left(\frac{\partial\rho^{(0)}}{\partial p^{(0)}}\right) = \left[\rho^{(0)}c^2\right]^{-1} , \tag{1.9.6}$$

which follows from (1.3.2). Here $K^{-1} = b$, the bulk modulus, which is used in Ch.3. The second term on the *rhs* of (1.9.3) is the kinetic energy density. As an analogy, we refer to the energy in a compressed spring. Here we have a compressible fluid, and the potential energy is the work done in compressing the fluid.

In terms of the velocity potential in (1.3.5) and (1.3.4) we can write (1.9.3) as

$$W = \frac{1}{2}K\left[\rho^{(0)}\right]^2(\partial_t\psi)^2 + \frac{1}{2}\rho^{(0)}(\partial_j\psi)^2 . \tag{1.9.7}$$

The Poynting vector and energy density are functions of time. If we take a time average of (1.9.1) we get

$$\frac{1}{T}\int_0^T \partial_t W dt + \frac{1}{T}\int_0^T \partial_j S_j dt = 0 . \tag{1.9.8}$$

Evaluating this we get

$$\frac{1}{T}\Big[W(t) - W(0)\Big] + \partial_j < S_j >= 0 , \tag{1.9.9}$$

where the bracket represents the temporal average. If $p^{(1)}$ and $v_j^{(1)}$ are periodic functions of time then $W(t)$ is also, and if the period is T then $W(T) = W(0)$ and

$$\partial_j < S_j >= 0 . \tag{1.9.10}$$

Further, using a harmonic representation of $p^{(1)}$ and $v_j^{(1)}$, what we really want are the real (Re) parts of these complex functions, i.e.

$$p^{(1)} = \text{Re}\left[p^{(1)}(\boldsymbol{x},t)\right] = \text{Re}\left[p^{(1)}(\boldsymbol{x})e^{-i\omega t}\right] \tag{1.9.11}$$

and

$$v_j^{(1)} = \text{Re}\left[v_j^{(1)}(\boldsymbol{x},t)\right] = \text{Re}\left[v_j^{(1)}(\boldsymbol{x})e^{-i\omega t}\right], \tag{1.9.12}$$

so that we have

$$\begin{aligned} < S_j > &= \frac{1}{T}\int_0^T S_j dt \\ &= \frac{1}{T}\int_0^T p^{(1)}v_j^{(1)} dt \\ &= \frac{1}{T}\int_0^T \text{Re}\left[p^{(1)}(\boldsymbol{x})e^{-i\omega t}\right]\text{Re}\left[v_j^{(1)}(\boldsymbol{x})e^{-i\omega t}\right] dt . \end{aligned}$$

Using real and imaginary parts of $p^{(1)}$ and $v_j^{(1)}$

$$p^{(1)} = p_r^{(1)} + ip_i^{(1)} \tag{1.9.13}$$

and

$$v_j^{(1)} = \left(v_j^{(1)}\right)_r + i\left(v_j^{(1)}\right)_i \ , \tag{1.9.14}$$

we can write the result as

$$< S_j >= \frac{1}{2}\mathrm{Re}\left[p^{(1)}(\boldsymbol{x})v_j^{(1)*}(\boldsymbol{x})\right] \ , \tag{1.9.15}$$

where the $*$ represents complex conjugation. The latter follows because we have

$$\int_0^T cos^2\omega t\, dt = \int_0^T \sin^2 \omega t\, dt = \frac{T}{2} \ ,$$

and

$$\int_0^T \sin\omega t \cos\omega t\, dt = 0 \ .$$

To find a relation involving the wave momentum we multiply (1.2.9) by $v_j^{(1)}$ and (1.2.8) by $p^{(1)}$ and add the resulting equations. The result can be written as

$$\partial_i t_{ij} + \partial_t m_j = 0 \ , \tag{1.9.16}$$

where

$$m_j = c^{-2}S_j = p^{(0)}v_j^{(1)} \ , \tag{1.9.17}$$

is the momentum density and

$$t_{ij} = \rho^{(0)}v_i^{(1)}v_j^{(1)} + \frac{1}{2}\delta_{ij}\left\{K\left[p^{(1)}\right]^2 - \rho^{(0)}\left[v_j^{(1)}\right]^2\right\} \ , \tag{1.9.18}$$

is called the wave stress tensor. Surface integrals of t_{ij} dotted into the surface normal n_j yield the tractions on a boundary.

1. Appendices

1.A Wave Equation Derivation for Spatially Varying Density

We derive (1.4.3) for the case when $\partial_i \rho^{(0)} \neq 0$ but $\partial_t \rho^{(0)} = 0$. The first consequence of this is that hydrostatic equilibrium doesn't hold so that we have

$$\partial_i p^{(0)} = \frac{\partial p^{(0)}}{\partial \rho^{(0)}} \partial_i \rho^{(0)} \ , \tag{1.A.1}$$

$$= c^2(\boldsymbol{x}) \partial_i \rho^{(0)} \ . \tag{1.A.2}$$

Since this is true there is a non-zero hydrodynamic velocity $v_j^{(0)}$. Substituting the resulting expansion (1.2.3) into (1.1.9) and equating coefficients of like order in ε we get (neglecting quadratic velocity terms)

$$\left[\rho^{(0)} + \varepsilon \rho^{(1)}\right] \left[\partial_t v_j^{(0)} + \varepsilon \partial_t v_j^{(1)}\right] = -\, \partial_j \left[p^{(0)} + \varepsilon p^{(1)}\right] \ ,$$

which to zeroth order is

$$\varepsilon^0 : \quad \rho^{(0)} \partial_t v_j^{(0)} = -\partial_j p^{(0)} \ , \tag{1.A.3}$$

and to first order is

$$\varepsilon : \quad \rho^{(0)} \partial_t v_j^{(1)} + \rho^{(1)} \partial_t v_j^{(0)} = -\partial_j p^{(1)} \ . \tag{1.A.4}$$

These equations replace (1.2.5) and (1.2.6). Using (1.2.7) and (1.A.3) the second term in (1.A.4) can be written as

$$\rho^{(1)} \partial_t v_j^{(0)} = -\frac{p^{(1)}}{c^2(\boldsymbol{x})} \frac{1}{\rho^{(0)}} \partial_j p^{(0)} \ ,$$

which simplifies further if we use (1.A.2) to write

$$\rho^{(1)}\partial_t v_j^{(0)} = -\frac{\partial_j \rho^{(0)}}{\rho^{(0)}} p^{(1)} \ . \tag{1.A.5}$$

Use of (1.A.5) in (1.A.4) yields

$$\begin{aligned} \partial_t v_j^{(1)} &= -\frac{1}{\rho^{(0)}}\partial_j p^{(1)} + \frac{1}{(\rho^{(0)})^2}\left[\partial_j \rho^{(0)}\right] p^{(1)} \\ &= -\partial_j \left\{\frac{1}{\rho^{(0)}} p^{(1)}\right\} \ . \end{aligned} \tag{1.A.6}$$

Next, substitute (1.2.7) into (1.2.4). The result replaces (1.2.9) and yields

$$\partial_t p^{(1)} + \rho^{(0)} c^2(\boldsymbol{x})\partial_i v_i^{(1)} + c^2(\boldsymbol{x}) v_i^{(1)} \partial_i \rho^{(0)} = 0 \ . \tag{1.A.7}$$

Differentiate (1.A.7) with respect to time and substitute from (1.A.6) to yield

$$\rho^{(0)}\partial_j \left\{\frac{1}{\rho^{(0)}}\partial_j p^{(1)}\right\} - \frac{1}{c^2(\boldsymbol{x})}\partial_t^2 p^{(1)} - M(\boldsymbol{x}) p^{(1)} = 0 \tag{1.A.8}$$

where

$$M(\boldsymbol{x}) = \frac{1}{\rho^{(0)}}\left\{\nabla^2 \rho^{(0)} - \frac{1}{\rho^{(0)}}\left[\partial_i \rho^{(0)}\right]^2\right\} \tag{1.A.9}$$

$$= \partial_i \left[\frac{1}{\rho^{(0)}}\partial_i \rho^{(0)}\right] \ . \tag{1.A.10}$$

This latter result is quoted in Sect. 1.4.

1.B Wave Equation Derivation for Spatially and Temporally Variable Density with Quadratic Velocity Terms

In this appendix we consider the derivation of the equations for pressure and velocity keeping the quadratic velocity terms. The conservation relations are (1.1.6):

$$\partial_t \rho + \partial_i(\rho v_i) = 0 \ ,$$

and (1.1.9)

$$\rho(\partial_t v_j + v_i \partial_i v_j) = -\, \partial_j p \ .$$

From (1.2.7) we have

$$p^{(1)} = \frac{\partial p^{(0)}}{\partial \rho^{(0)}} \rho^{(1)} = c^2(\boldsymbol{x}, t) \rho^{(1)} \ . \tag{1.B.1}$$

Also, if $p^{(0)} = p^{(0)}(\rho^{(0)})$,

$$\partial_j p^{(0)} = \frac{\partial p^{(0)}}{\partial \rho^{(0)}} \partial_j \rho^{(0)} = c^2(\boldsymbol{x}, t) \partial_j \rho^{(0)} \ . \tag{1.B.2}$$

Substitute (1.2.3) into (1.1.6):

$$\partial_t \left[\rho^{(0)} + \varepsilon \rho^{(1)}\right] + \partial_i \Big\{ \left(\rho^{(0)} + \varepsilon \rho^{(1)}\right) \left(v_i^{(0)} + \varepsilon v_i^{(1)}\right) \Big\} = 0 \ ,$$

and equate coefficients of like powers to yield to zeroth order

$$\varepsilon^0 : \quad \partial_t \rho^{(0)} + \partial_i (\rho^{(0)} v_i^{(0)}) = 0 \ , \tag{1.B.3}$$

and to first order

$$\varepsilon : \quad \partial_t \rho^{(1)} + \partial_i (\rho^{(1)} v_i^{(0)} + \rho^{(0)} v_i^{(1)}) = 0 \ ,$$

or

$$\partial_t \rho^{(1)} + v_i^{(0)} \partial_i \rho^{(1)} + \rho^{(1)} \partial_i v_i^{(0)} + \partial_i \left(\rho^{(0)} v_i^{(1)}\right) = 0 \ . \tag{1.B.4}$$

From (1.B.3)

$$\partial_i v_i^{(0)} = -\, R_i v_i^{(0)} - R_t \ , \tag{1.B.5}$$

where R_i is defined by

$$R_i = \frac{1}{\rho^{(0)}} \partial_i \rho^{(0)} \ , \tag{1.B.6}$$

and R_t by

$$R_t = \frac{1}{\rho^{(0)}} \partial_t \rho^{(0)} \ . \tag{1.B.7}$$

Then we can remove the divergence in (1.B.4)

$$\partial_t \rho^{(1)} + v_i^{(0)} \left[\partial_i \rho^{(1)} - R_i \rho^{(1)}\right] - \rho^{(1)} R_t + \partial_i \left[\rho^{(0)} v_i^{(1)}\right] = 0 \ . \tag{1.B.8}$$

Substitute for $\rho^{(1)}$ from (1.B.1) and define

$$R(\boldsymbol{x}, t) = v_i^{(0)} R_i + R_t \ , \tag{1.B.9}$$

so that

$$\partial_t \left[\frac{p^{(1)}}{c^2}\right] - \frac{p^{(1)}}{c^2} R + v_i^{(0)} \partial_i \left[\frac{p^{(1)}}{c^2}\right] + \partial_i \left[\rho^{(0)} v_i^{(1)}\right] = 0 \ . \quad (1.B.10)$$

Since $p^{(1)}/c^2$ will occur here and in the next field equation, define

$$P = \frac{p^{(1)}}{c^2} \ , \quad (1.B.11)$$

and the operator L (Lagrangian derivative)

$$L = \partial_t + v_i^{(0)} \partial_i \ , \quad (1.B.12)$$

to give the *first field equation*

$$LP - RP + \partial_i \left[\rho^{(0)} v_i^{(1)}\right] = 0 \ . \quad (1.B.13)$$

Next, substitute (1.2.3) into (1.1.9) to yield:

$$(\rho^{(0)} + \varepsilon \rho^{(1)}) \Big\{ \partial_t v_j^{(0)} + \varepsilon \partial_t v_j^{(1)} + \left(v_i^{(0)} + \varepsilon v_i^{(1)}\right) \partial_i \left(v_j^{(0)} + \varepsilon v_j^{(1)}\right) \Big\}$$
$$= - \partial_j p^{(0)} - \varepsilon \partial_j p^{(1)} \ , \quad (1.B.14)$$

and again equate coefficients. To zeroth order we get

$$\varepsilon^0 : \quad \rho^{(0)} (\partial_t v_j^{(0)} + v_i^{(0)} \partial_i v_j^{(0)}) = - \partial_j p^{(0)} \ , \quad (1.B.15)$$

and to first order

$$\varepsilon : \ \rho^{(1)} \left(\partial_t v_j^{(0)} + v_i^{(0)} \partial_i v_j^{(0)}\right)$$
$$+ \rho^{(0)} \left(\partial_t v_j^{(1)} + v_i^{(0)} \partial_i v_j^{(1)} + v_i^{(1)} \partial_i v_j^{(0)}\right) = - \partial_j p^{(1)} \ . \quad (1.B.16)$$

Multiply (1.B.16) by $\rho^{(0)}$. Use (1.B.15) and (1.B.1) to yield

$$\frac{p^{(1)}}{c^2} (- \partial_j p^{(0)}) + (\rho^{(0)})^2 \Big(\partial_t v_j^{(1)} + v_i^{(0)} \partial_i v_j^{(1)} + v_i^{(1)} \partial_i v_j^{(0)} \Big) = - \rho^{(0)} \partial_j p^{(1)} \ . \quad (1.B.17)$$

Use (1.B.2) and (1.B.12) to derive

$$- \frac{p^{(1)}}{c^2} (c^2 \partial_j \rho^{(0)}) + (\rho^{(0)})^2 \Big(L v_j^{(1)} + v_i^{(1)} \partial_i v_j^{(0)} \Big) = - \rho^{(0)} \partial_j p^{(1)} \ . \quad (1.B.18)$$

Use:

$$\begin{aligned}
\partial_j p^{(1)} &= \partial_j \left[\frac{p^{(1)}}{c^2} c^2 \right] \\
&= c^2 \partial_j \left[\frac{p^{(1)}}{c^2} \right] + \frac{p^{(1)}}{c^2} \partial_j c^2 && (1.B.19) \\
&= c^2 (\partial_j P + C_j P) && (1.B.20)
\end{aligned}$$

where we have used (1.B.11) and the definition

$$C_j = \frac{1}{c^2} \partial_j c^2 = \frac{2}{c} \partial_j c \ . \qquad (1.B.21)$$

Use (1.B.6) and (1.B.18) to give the *second field equation*

$$\partial_j P + (C_j - R_j) P + \frac{\rho^{(0)}}{c^2} \left(L v_j^{(1)} + v_i^{(1)} \partial_i v_j^{(0)} \right) = 0 \ . \qquad (1.B.22)$$

Using the two field equations (1.B.13) and (1.B.22) we can derive wave equations on pressure or velocity proceeding as in the derivation for constant density in the chapter. See for example Refs.[1.4] and [1.9].

2. Green's Functions

In this chapter we derive Green's functions useful for scattering and propagation problems. We present a thorough discussion of one-dimensional problems as well as many examples. The Green's functions for the wave and Helmholtz equations are derived and interrelated for one, two, and three dimensions. Useful integral representations of these functions are presented as well as the relation between delta functions in different coordinate systems.

2.1 Basic Definition

The differential equations for the S-L problems in Ch. 1 contained no forcing term, i.e. their right-hand sides were set to zero. Here we define a one-dimensional equation with a delta function on the rhs, viz.

$$\frac{d}{dz}\left[p(z)\frac{dG}{dz}\right] + \left[q(z) + \lambda r(z)\right]G = -\delta(z - z_0) \ . \tag{2.1.1}$$

The solution of this equation, $G(z, z_0; \lambda)$, is the one-dimensional Green's function. It will satisfy appropriate boundary conditions discussed below.

Recall that $\delta(z - z_0)$ vanishes if $z \neq z_0$ and

$$\int_{z_1}^{z_2} \delta(z - z_0)dz = \begin{cases} 1 & z_0 \in [z_1, z_2] \ , \\ 0 & z_0 \notin [z_1, z_2] \ . \end{cases} \tag{2.1.2}$$

Assume that p', q and r are continuous functions, and make the following assumptions on G:

1. G is continuous everywhere in the interval of definition.

2. dG/dz is continuous everywhere in the interval of definition except at $z = z_0$.

Next, integrate (2.1.1) from $z_0 - \varepsilon$ to $z_0 + \varepsilon$ where $\varepsilon > 0$ and we let it approach zero. We have the following results

$$\int_{z_0-\varepsilon}^{z_0+\varepsilon} \delta(z - z_0)dz = 1 \ , \tag{2.1.3}$$

$$\lim_{\varepsilon \to 0} \int_{z_0-\varepsilon}^{z_0+\varepsilon} \Big[q(z) + \lambda r(z)\Big] G(z, z_0; \lambda)dz = 0 \ , \tag{2.1.4}$$

which is true if q, r and G are all continuous (it actually holds if q and r are only piecewise continuous), and

$$\begin{aligned}\int_{z_0-\varepsilon}^{z_0+\varepsilon} \frac{d}{dz}\left[p(z)\frac{dG}{dz}\right] dz \ &= \ p(z_0 + \varepsilon)\frac{dG}{dz}(z_0 + \varepsilon,\ z_0; \lambda) \\ &\quad - p(z_0 - \varepsilon)\frac{dG}{dz}(z_0 - \varepsilon,\ z_0;\ \lambda) \ .\end{aligned} \tag{2.1.5}$$

If we further assume that p is continuous, and define upper (U) and lower (L) solutions as

$$\lim_{\varepsilon \to 0} \frac{dG}{dz}(z_0 \pm \varepsilon,\ z_0;\ \lambda) = \begin{cases} \dfrac{dG^U}{dz}(z_0,\ z_0;\ \lambda) \ , \\[2ex] \dfrac{dG^L}{dz}(z_0,\ z_0;\ \lambda) \ , \end{cases} \tag{2.1.6}$$

the result of integrating (2.1.1) becomes

$$\frac{dG^U}{dz}(z_0,\ z_0;\ \lambda) - \frac{dG^L}{dz}(z_0,\ z_0;\ \lambda) = -\frac{1}{p(z_0)} \ , \tag{2.1.7}$$

which specifically defines the discontinuity of the derivative of G at $z = z_0$. We use (2.1.7) with the continuity of G at z_0,

$$G^U(z_0,\ z_0;\ \lambda) - G^L(z_0,\ z_0;\ \lambda) = 0 \ , \tag{2.1.8}$$

to determine the full function. Thus, the delta function creates an artificial layer or interface for our problem, and (2.1.7) and (2.1.8) specify how the solution crosses this layer. Two conditions are necessary since we have a second-order equation.

2.2 Eigenfunction Expansion

To solve (2.1.1), we begin with the eigenfunction solutions $\psi_n(z)$, of the homogeneous equation

$$\frac{d}{dz}\left[p(z)\frac{d\psi_n}{dz}\right] + \left[q(z) + \lambda_n r(z)\right]\psi_n = 0 \ . \tag{2.2.1}$$

For a given boundary value problem these functions are assumed known. Write G as a series of these eigenfunctions

$$G(z,\ z_0;\ \lambda) = \sum_n A_n \psi_n(z) \ , \tag{2.2.2}$$

where the A_n are constants to be determined, and substitute (2.2.2) in (2.1.1). Interchanging summation and differentiation, and substituting (2.2.1) yields

$$r(z) \sum_n A_n \left(\lambda - \lambda_n\right) \psi_n(z) = \delta\left(z - z_0\right) \ . \tag{2.2.3}$$

Multiply (2.2.3) by $\psi_m(z)$, integrate from a to b and use the orthogonality of the eigenfunctions to yield (we assume real eigenfunctions for simplicity)

$$A_m(\lambda_m - \lambda) = \psi_m(z_0) \ , \tag{2.2.4}$$

so that (2.2.2) can be written as

$$G(z, z_0; \lambda) = \sum_n \frac{\psi_n(z)\psi_n(z_0)}{\lambda_n - \lambda} \ . \tag{2.2.5}$$

This is a solution for G in terms of the eigenvalues λ_n and eigenfunctions $\psi_n(z)$ of the homogeneous problem. Note that G is a symmetric function of source-receiver positions

$$G(z, z_0; \lambda) = G(z_0, z; \lambda) \ , \tag{2.2.6}$$

and further note that the ψ_n depend on λ_n but *not* on λ.

Up to now we have considered S-L problems with all real quantities. Now we digress and assume that λ is a complex variable. The representation (2.2.5) indicates that G as a function of λ has only simple poles in the λ-plane at values λ_n. Define the contour C_λ as a counter-clockwise closed contour in the λ-plane enclosing all the singularities of G. Then if we integrate (2.2.5) around this contour we get

$$\int_{C_\lambda} G(z,\ z_0;\ \lambda) d\lambda = -2\pi i \sum_n \psi_n(z)\psi_n(z_0) \ , \tag{2.2.7}$$

which because of the completeness relation (1.6.26) becomes

$$\frac{1}{2\pi i}\int_{C_\lambda} G(z,\ z_0;\ \lambda) d\lambda = -\frac{\delta(z - z_0)}{r(z)} \ . \tag{2.2.8}$$

Contour integrals of this type will be used in Chs. 4 and 5.

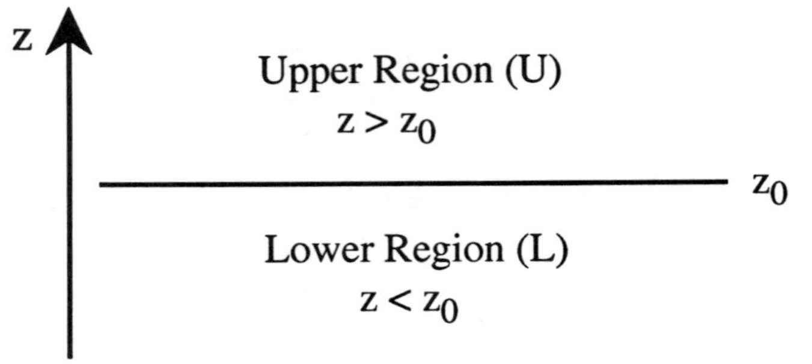

Fig. 2.1. Geometry for the infinite space example.

2.3 One-Dimensional Examples

We consider two examples of this development, one whose "boundaries" are at infinity, and the second with finite boundaries (a waveguide). For simplicity we set $p = 1$ in (2.1.1) so that G satisfies

$$\frac{d^2}{dz^2}G + \Big[q(z) + \lambda r(z)\Big] G(z,\, z_0;\, \lambda) = -\,\delta(z - z_0)\ . \tag{2.3.1}$$

The eigenfunctions will then satisfy the equation

$$\frac{d^2}{dz^2}\psi + \Big[q(z) + \lambda r(z)\Big]\psi = 0\ , \tag{2.3.2}$$

which has two linearly independent solutions ψ_1 and ψ_2.

Example 1: Infinite Space

For this case we actually have two semi-infinite half spaces joined at the artificial interface created at the source point z_0 of the delta function.[1] See Fig.2.1. To solve for G, we first suppose that we are in the upper region $z > z_0$. We have linearly independent solutions ψ_1 and ψ_2, and the upper (U) solution for G could be written in several possible ways, viz.

$$G^U(z, z_0; \lambda) = \begin{cases} \psi_1\ , \\ \psi_2\ , \\ \psi_1 + \alpha\psi_2\ . \end{cases} \tag{2.3.3}$$

[1] For simplicity we assume that G and ψ satisfy the same equation in both half-spaces.

The choice of the solution depends on the region. In our case, z could approach $+\infty$. At $+\infty$ there is a physical boundary condition in that the solution should not blow up. In particular, since we are interested in describing wave phenomena, we assume that as $z \to +\infty$, G^U should behave like an outgoing wave or a decaying solution. Growing solutions are unphysical, and incoming waves must have a source, a case which we have excluded by restricting the only source point to z_0 and eliminating any boundaries. To completely exclude incoming waves, we must also say something about the behavior of $q(z)$ and $r(z)$ which we treat below. In particular, in one dimension if

$$G^U(z, z_0; \lambda) \sim e^{ikz} \quad , \quad \mathrm{Im}\, k \geq 0 \ , \tag{2.3.4}$$

(i.e. if the upper solution behaved like a plane wave) this would satisfy our physical boundary condition. But we have a homogeneous equation (2.3.2) where $q(z)$ and $r(z)$ are in general complicated functions. So the question arises of how to get a plane wave. It is possible to get it if as $z \to +\infty$

$$q(z) \to 0; \ r(z) \to 1; \ \lambda = k^2 \ , \tag{2.3.5}$$

where q and r must approach their limits monotonically. (If they don't, for example if one of them has a kink, the latter region can serve as a source of trapped waves and correspondingly a source for incoming waves between the kink and z_0). We then pick that combination of solutions in (2.3.3) which approaches a plane wave $\exp(ikz)$ as $z \to +\infty$. For simplicity, assume its ψ_1, i.e.

$$G^U(z, z_0; \lambda) = A\, \psi_1(z) \quad , \quad z > z_0 \ . \tag{2.3.6}$$

where the constant A must be determined.

Similarly for $z < z_0$, we assume that (2.3.5) is also true as $z \to -\infty$ and that the lower solution behaves as an outgoing wave $\exp(-ikz)$ as $z \to -\infty$. For simplicity we assume this solution is ψ_2, so that

$$G^L(z, z_0; \lambda) = B\, \psi_2(z) \quad , \quad z < z_0 \ , \tag{2.3.7}$$

where the constant B must be determined.

We determine A and B by substituting (2.3.6) and (2.3.7) in (2.1.7) and (2.1.8), which become

$$A\psi_1'(z_0) - B\psi_2'(z_0) = -1 \ ,$$

and

$$A\, \psi_1(z_0) - B\, \psi_2(z_0) = 0 \ . \tag{2.3.8}$$

These have the solution

$$A = \psi_2(z_0)W^{-1}; \; B = \psi_1(z_0)W^{-1} \quad , \tag{2.3.9}$$

where W is the Wronskian defined in (1.6.15). The full Green's function is thus

$$G(z,\, z_0;\, \lambda) = \begin{cases} \psi_1(z)\psi_2(z_0)W^{-1} \; , & z > z_0 \; , \\ \psi_1(z_0)\psi_2(z)W^{-1} \; , & z < z_0 \; . \end{cases} \tag{2.3.10}$$

As a specific example, let $q = 0$, $r = 1$ and $\lambda = k^2$ so that

$$\frac{d^2}{dz^2}\psi + k^2\psi = 0 \quad , \tag{2.3.11}$$

(so the equation is the limiting case of the equation we described) and

$$\psi_1(z) = \exp(ikz); \quad \psi_2(z) = \exp(-ikz); \quad W = -2ik \quad , \tag{2.3.12}$$

and thus

$$G(z,\, z_0;\, k^2) = (-2ik)^{-1}\exp(ik|z - z_0|) \quad . \tag{2.3.13}$$

This function is just the one-dimensional free-space Green's function which satisfies

$$\frac{d^2}{dz^2}G + k^2G = -\delta(z - z_0) \quad . \tag{2.3.14}$$

We return to it later in Sect. 2.8. Considered as a function of λ, (2.3.13) contains a branch point at $\lambda = 0$. The contour C_λ in (2.2.7) encircles a branch cut and the function has a continuum of values of λ as possible eigenvalues. That is, the boundary value problem in infinite space has a continuous spectrum of eigenvalues. This is analogous to the continuous spectrum possessed by a free particle in quantum mechanics (see Ref. 1.1).

Example 2: Two layer Bounded

Here we again want to solve (2.3.1) but now we introduce upper (z_U) and lower (z_L) boundaries which we assume to be perfectly reflecting. See Fig.2.2. The net physical effect is to trap waves in the region $z_L \le z \le z_U$ and this will lead to standing wave solutions. The solution of (2.3.1) can be written as

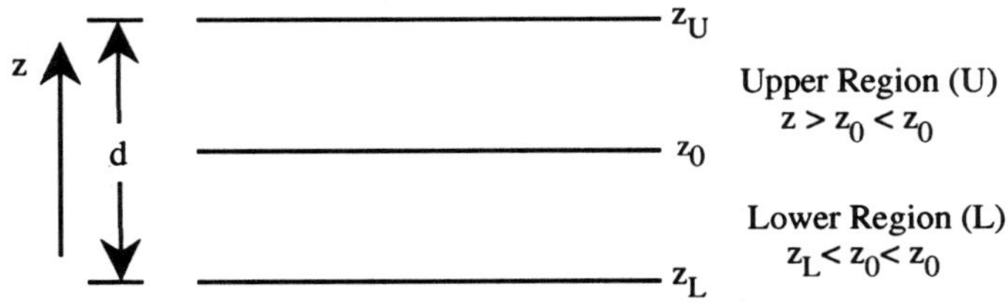

Fig. 2.2. Geometry for the two layer bounded example of width $d = z_U - z_L$.

$$G(z,\, z_0;\, \lambda) = \begin{cases} G^U(z,\, z_0;\, \lambda)\,, & z_0 \le z \le z_U\ , \\ G^L(z, z_0; \lambda)\,, & z_L \le z \le z_0\ . \end{cases} \tag{2.3.15}$$

The upper and lower solutions are each written in terms of the linearly independent solutions of the homogeneous problem (2.3.2),

$$G^U(z,\, z_0;\, \lambda) = A\,\psi_1(z) + B\,\psi_2(z)\ , \tag{2.3.16}$$

and

$$G^L(z,\, z_0;\, \lambda) = C\psi_1(z) + D\psi_2(z)\ , \tag{2.3.17}$$

where the four constants A, B, C, and D are unknown. One of the constants A and B is determined from the boundary condition on G^U at $z = z_U$. (This replaces the outgoing wave condition in Example 1). One of the constants C and D is found from the boundary condition on G^L at $z = z_L$, replacing the outgoing wave boundary condition as $z \to -\infty$ in Example 1. The remaining two constants are found from the jump condition on the derivative (2.1.7) at $z = z_0$ which is

$$\frac{d}{dz}\,G^U(z_0,\, z_0;\, \lambda) - \frac{d}{dz}\,G^L(z_0,\, z_0;\, \lambda) = -1\ , \tag{2.3.18}$$

and the continuity condition (2.1.8) which is

$$G^U(z_0,\, z_0;\, \lambda) - G^L(z_0,\, z_0;\, \lambda) = 0\ . \tag{2.3.19}$$

As a specific example, we again assume that $q = 0$, $r = 1$ and $\lambda = k^2$ so that G satisfies (2.3.14) and ψ_1 and ψ_2 satisfy (2.3.11). Example 1 was solved in free space. Here we choose the boundary conditions

$$G^U(z_U,\, z_0;\, k^2) = 0\ , \tag{2.3.20}$$

and

$$\frac{d}{dz} G^L(z_L,\, z_0;\, k^2) = 0 \ . \tag{2.3.21}$$

To find the solution G^U, we note that the linearly independent solutions ψ_1 and ψ_2 can be chosen in various ways. The two simplest are $\exp(\pm ikz)$ and $(\sin kz,\, \cos kz)$. These can be either directly inserted in (2.3.16) and the boundary condition (2.3.20) applied, or we can choose the solutions ψ_1 and ψ_2 in the upper region as

$$\psi_1^U(z) = \sin\Big[k(z - z_U)\Big] \ , \tag{2.3.22}$$

and

$$\psi_2^U(z) = \cos\Big[k(z - z_U)\Big] \ , \tag{2.3.23}$$

and note that the application of the boundary condition (2.3.20) immediately yields $B = 0$, so that we have

$$G^U(z,\, z_0;\, k^2) = A\,\psi_1^U(z) \ . \tag{2.3.24}$$

Similarly, in the lower region we can choose either ψ_1 and ψ_2 as $\exp(\pm ikz)$ or the pair $(\sin kz,\, \cos kz)$ and apply the boundary condition (2.3.21). Also note that if we choose

$$\psi_1^L(z) = \sin\Big[k(z - z_L)\Big] \ , \tag{2.3.25}$$

and

$$\psi_2^L(z) = \cos\Big[k(z - z_L)\Big] \ , \tag{2.3.26}$$

and apply (2.3.21), we immediately get $C = 0$ and thus

$$G^L(z,\, z_0;\, k^2) = D\psi_2^L(z) \ . \tag{2.3.27}$$

The constants A and D are determined from (2.3.18) and (2.3.19) to be

$$A = \psi_2^L(z_0)W^{-1};\ D = \psi_1^U(z_0)W^{-1} \ , \tag{2.3.28}$$

where W is the Wronskian. The full solution (2.3.15) is thus given by

$$G(z,\, z_0;\, k^2) = \begin{cases} \psi_1^U(z)\psi_2^L(z_0)W^{-1} & z_0 \le z \le z_U \ , \\ \psi_1^U(z_0)\psi_2^L(z)W^{-1} & z_L \le z \le z_0 \ , \end{cases} \tag{2.3.29}$$

and the explicit evaluation of the Wronskian is

$$W = \frac{d}{dz}\psi_2^L(z_0)\psi_1^U(z_0) - \psi_2^L(z_0)\frac{d}{dz}\psi_1^U(z_0)$$

$$= -k\cos(kd)\ , \quad d = z_U - z_L\ . \tag{2.3.30}$$

We can thus write (2.3.29) as

$$G(z, z_0; k^2) = \begin{cases} \sin\big[k(z_U - z)\big]\cos\big[k(z_0 - z_L)\big]/k\cos(kd) : z_0 \le z \le z_U\ , \\ \sin\big[k(z_U - z_0)\big]\cos\big[k(z - z_L)\big]/k\cos(kd) : z_L \le z \le z_0\ . \end{cases} \tag{2.3.31}$$

Note that W depends parametrically on $k = \sqrt{\lambda}$, and that as a function of k, G has simple poles at those points where the Wronskian vanishes, provided the numerator doesn't also vanish. These occur when

$$\cos(kd) = 0\ , \tag{2.3.32}$$

or for those values

$$\sqrt{\lambda_n} = k_n = (n + 1/2)\pi d^{-1}\ , \tag{2.3.33}$$

where $n = 0,\ \pm 1,\ \pm 2, \cdots$. The fact of simple poles can be illustrated explicitly using the infinite product representation of the cosine

$$\cos(kd) = \prod_{n=0}^{\infty}\left\{1 - \frac{4k^2d^2}{(2n+1)^2\pi^2}\right\}\ . \tag{2.3.34}$$

The bounded region problem thus possesses a purely discrete spectrum.

2.4 Elementary Source

The Green's function is an elementary source solution in the following sense. The field $F(z)$ which results from an arbitrary source $S(z)$ satisfies the differential equation

$$\frac{d}{dz}\left[p(z)\frac{dF}{dz}\right] + \big[q(z) + \lambda r(z)\big]F = -S(z)\ . \tag{2.4.1}$$

It can be synthesized from the elementary source solution $G(z,\ z_0;\ \lambda)$ which satisfies

$$\frac{d}{dz}\left[p(z)\frac{dG}{dz}\right] + \left[q(z) + \lambda r(z)\right]G = -\delta(z - z_0) \ , \qquad (2.4.2)$$

by the integral

$$F(z) = \int G(z,\ z_0;\ \lambda)S(z_0)dz_0 \ , \qquad (2.4.3)$$

where the integration is over the region of extent of the source S. It can be seen that F is a solution of (2.4.1) by substituting (2.4.3) into (2.4.1) and using (2.4.2). The differentiation is in terms of the field variable z. Note also that F given by (2.4.3) satisfies the same boundary conditions as G.

2.5 Green's Functions for the Wave Equation

In this section we solve for the Green's function in three spatial and one temporal dimension, i.e.

$$G^{(3,1)}(\boldsymbol{x},\ t; \boldsymbol{x}',\ t') \ , \qquad (2.5.1)$$

where $\boldsymbol{x}'$, t' are the position and time at the source, and $\boldsymbol{x}$, t the position and time at the receiver. The function satisfies the wave equation

$$(\nabla_x^2 - c^{-2}\partial_t^2)G^{(3,1)}(\boldsymbol{x},\ t;\ \boldsymbol{x}',\ t') = -\delta(\boldsymbol{x} - \boldsymbol{x}')\delta(t - t') \ , \qquad (2.5.2)$$

where the Laplacian is

$$\nabla_x^2 = \partial_x^2 + \partial_y^2 + \partial_z^2 \ . \qquad (2.5.3)$$

Once we find this function, there is a systematic way to find the corresponding Green's functions for wave equations in other spatial dimensions, and Green's functions for the steady state Helmholtz equation. We discuss these in later sections.

We treat the solution of this equation in two ways. The first, using four-vector notation, is to elicit the more general properties of the function. The second, using Fourier transforms, is to connect our results to the one-dimensional Green's functions.

We first introduce the four-vector notation

$$x = (\boldsymbol{x},\ x_o) \ , \quad x_0 = ct \ , \qquad (2.5.4)$$

and the scalar product

$$x \cdot x = \boldsymbol{x} \cdot \boldsymbol{x} - x_0^2 \begin{cases} > 0 & \text{space-like}\,, \\ = 0 & \text{light-like}\,, \\ < 0 & \text{time-like}\,. \end{cases} \tag{2.5.5}$$

The Green's function $G^{(3,1)}(x, x')$ thus satisfies

$$\begin{aligned} \Box G^{(3,1)}(x, x') &= -\delta(x - x') \\ &= -\delta(\boldsymbol{x} - \boldsymbol{x}')\delta(x_0 - x_0') \\ &= -\delta(\boldsymbol{x} - \boldsymbol{x}')c^{-1}\delta(t - t')\ , \end{aligned} \tag{2.5.6}$$

where we have defined the d'Alembert operator

$$\Box = \nabla_{\boldsymbol{x}}^2 - \partial_0^2\ , \tag{2.5.7}$$

with

$$\partial_0 = c^{-1}\partial_t\ . \tag{2.5.8}$$

The function defined by (2.5.6) is related to the one defined in (2.5.2) by

$$cG^{(3,1)}(x, x') = G^{(3,1)}(\boldsymbol{x},\, t;\, \boldsymbol{x}',\, t')\ . \tag{2.5.9}$$

We note that this function is translationally invariant,

$$G^{(3,1)}(x, x') = G^{(3,1)}(x - x')\ , \tag{2.5.10}$$

i.e. it only depends on the difference of its arguments. Further, it is the Green's function for an elementary source, since the field F due to an arbitrary source $S(x)$ which has both spatial and temporal extent can be written as ($d^4x' = d\boldsymbol{x}'dx_0'$)

$$F(x) = \int\!\!\int\!\!\int\!\!\int G^{(3,1)}(x, x')S(x')d^4x'\ , \tag{2.5.11}$$

and satisfies the equation

$$\Box F(x) = -S(x)\ . \tag{2.5.12}$$

Since this is only a function of the difference argument we introduce the Fourier transform as

$$G^{(3,1)}(x) = (2\pi)^{-4}\int\!\!\int\!\!\int\!\!\int e^{ik\cdot x}\tilde{G}^{(3,1)}(k)d^4k\ . \tag{2.5.13}$$

Then ($k_0 = \omega c^{-1}$)

$$\Box G^{(3,1)}(x) = (2\pi)^{-4} \int\!\!\int\!\!\int\!\!\int (-\boldsymbol{k}^2 + k_0^2) e^{ik\cdot x} \tilde{G}^{(3,1)}(k) d^4k \ . \quad (2.5.14)$$

If we choose

$$\tilde{G}^{(3,1)}(k) = (\boldsymbol{k}^2 - k_0^2)^{-1} = (k^2)^{-1} \ , \quad (2.5.15)$$

then

$$\Box G^{(3,1)}(x) = -\delta(x) \ , \quad (2.5.16)$$

where

$$\delta(x) = (2\pi)^{-4} \int\!\!\int\!\!\int\!\!\int e^{ik\cdot x} d^4k \ , \quad (2.5.17)$$

so that (2.5.13) can be written as

$$G^{(3,1)}(x) = -(2\pi)^{-4} \int\!\!\int\!\!\int\!\!\int \frac{e^{ik\cdot x}}{k_0^2 - W^2} d^4k \ , \quad (2.5.18)$$

where

$$W = |\boldsymbol{k}| \ . \quad (2.5.19)$$

As it stands, however, (2.5.18) is not well defined since, for example, the k_0-integral is singular. We must, thus, define how to treat integrals of the form

$$I = \int_{-\infty}^{\infty} \frac{e^{-ik_0x_0}}{(k_0 - W)(k_0 + W)} dk_0 \ , \quad (2.5.20)$$

with singularities in the complex k_0-plane at $\pm W$ (see Fig.2.3). There are two ways to handle this. We can either fix the poles and offset the contour illustrated in the left hand sketches in Fig.2.3 parts (b), (c) and (d), or fix the contour and offset the poles as in the right hand sketches in Fig.2.3. We choose the latter, i.e. we shift the poles an amount $\varepsilon > 0$ where we assume $\varepsilon \to 0$. The poles in the integrand of I in (2.5.20) are thus rewritten as

$$(k_0 - W)(k_0 + W) \to \Big[k_0 - (W + i\alpha\varepsilon)\Big]\Big[k_0 + W - i\beta\varepsilon\Big] \ , \quad (2.5.21)$$

where α and β can equal $0, \pm 1$ depending on the way they are shifted. The integral I thus becomes

$$I(\alpha,\beta) = \int_{-\infty}^{\infty} \frac{e^{-ik_0x_0}}{(k_0 - W - i\alpha\varepsilon)(k_0 + W - i\beta\varepsilon)} dk_0 \ . \quad (2.5.22)$$

To evaluate the poles we can use the Dirac Plemelj relations (derived in Appendix 2.A)

$$\frac{1}{y \pm i\varepsilon} = P\frac{1}{y} \mp \pi i \delta(y) \ , \quad (2.5.23)$$

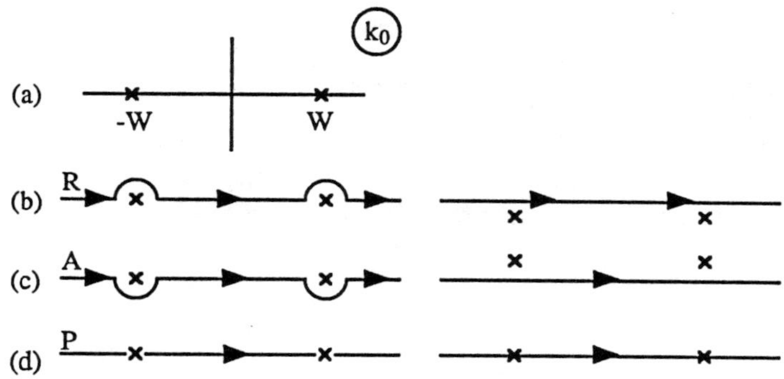

Fig. 2.3. (a) Poles of the integrand of I in the complex k_0-plane. Three ways of fixing the poles/offsetting the contour or fixing the contour/offsetting the poles for (b) retarded (R), (c) advanced (A) and (d) principal value (P) Green's functions.

where y is a scalar and P represents the Cauchy principal value distribution. For example this yields

$$\frac{1}{k_0 - W - i\alpha\varepsilon} = P\frac{1}{k_0 - W} + \alpha\pi i\delta(k_0 - W) \ , \qquad (2.5.24)$$

and

$$\frac{1}{(k_0 - W - i\alpha\varepsilon)(k_0 + W - i\beta\varepsilon)} = P\frac{1}{k_0^2 - W^2} + \frac{\pi i}{2W}\Big\{\alpha\delta(k_0 - W) - \beta\delta(k_0 + W)\Big\} \ . \qquad (2.5.25)$$

Alternatively we can evaluate I using conventional means.

We first define a function by shifting both poles below the axis as in Fig.2.3b for which we have $\alpha = -1$ and $\beta = -1$. This is referred to as the Retarded (R) Green's Function for reasons we explain below. We have that

$$G_R^{(3,1)}(x) = -(2\pi)^{-4}\int\int\int e^{i\boldsymbol{k}\cdot\boldsymbol{x}} I(-1,-1) d\boldsymbol{k} \ , \qquad (2.5.26)$$

where

$$I(-1,-1) = \int_{-\infty}^{\infty} \frac{e^{-ik_0 x_0}}{(k_0 - W + i\varepsilon)(k_0 + W + i\varepsilon)}\, dk_0 \ . \qquad (2.5.27)$$

This latter integral can be evaluated using residue calculus methods. In the complex k_0-plane both poles are in the lower half plane. Closing the contour in the upper half plane ($x_0 < 0$) with a semi-circle whose radius approaches infinity, we find no pole contributions, and the contribution from the semicircular contour vanishes by Jordan's Lemma. Closing in the lower half plane ($x_0 > 0$) yields the two residue contributions and again the contribution from the semi-circular contour vanishes by Jordan's lemma. The result is

$$I(-1,-1) = (\pi i/W)\left(e^{ix_0 W} - e^{-ix_0 W}\right)\theta(x_0) \ , \tag{2.5.28}$$

with the step function

$$\theta(x_0) = \begin{cases} 1 & x_0 > 0 \ , \\ 0 & x_0 < 0 \ . \end{cases} \tag{2.5.29}$$

The remaining three integrals in (2.5.26) can be evaluated using spherical polar coordinates. Define

$$\begin{aligned} \boldsymbol{k}\cdot\boldsymbol{x} &= |\boldsymbol{k}|\,|\boldsymbol{x}|\cos\theta = Wr\cos\theta \ , \\ d\boldsymbol{k} &= W^2 dW\Big[-d\cos\theta\Big]d\phi \ , \end{aligned} \tag{2.5.30}$$

where $\theta : 0 \to \pi$ and $\phi : 0 \to 2\pi$. The ϕ-integration just yields 2π and (2.5.26) with (2.5.28) becomes

$$\begin{aligned} G_R^{(3,1)}(x) &= -(2\pi)^{-4} 2\pi^2 i\theta(x_0) \\ &\cdot \int_0^\infty W(e^{ix_0 W} - e^{-ix_0 W})dW \int_0^\pi e^{iWr\cos\theta}(-d\cos\theta) \ . \end{aligned} \tag{2.5.31}$$

The θ-integral is an exact differential and the resulting integrand in W is an even function of W. Evaluating this we get

$$G_R^{(3,1)}(x) = -\frac{\theta(x_0)}{4\pi r}\Big[\delta(x_0 + r) - \delta(r - x_0)\Big] \ . \tag{2.5.32}$$

Since $r \geq 0$, we have that

$$\theta(x_0)\delta(x_0 + r) = 0 \ , \tag{2.5.33}$$

so that

$$\begin{aligned} G_R^{(3,1)}(x) &= \frac{\theta(x_0)}{4\pi r}\delta(x_0 - r) \\ &= \frac{\theta(x_0)}{4\pi rc}\delta(t - r/c) \ , \end{aligned} \tag{2.5.34}$$

and

$$G_R^{(3,1)}(\boldsymbol{x}, t) = \frac{\theta(t)}{4\pi r}\delta(t - r/c) \ . \tag{2.5.35}$$

(The step functions in (2.5.34) and (2.5.35) are often considered redundant since the delta functions only have positive temporal support). Because of the translational invariance we have that

$$G_R^{(3,1)}(\boldsymbol{x}, t; \boldsymbol{x}', t') = \frac{\theta(t - t')}{4\pi|\boldsymbol{x} - \boldsymbol{x}'|}\delta(t - t' - |\boldsymbol{x} - \boldsymbol{x}'|/c) \ . \tag{2.5.36}$$

Equations (2.5.34) or (2.5.35) say the following about $G_R^{(3,1)}$:

1. It is a causal function, i.e. it vanishes for $t < 0$.

2. At $t = 0$ it begins as a point source pulse. After time t has elapsed the pulse is concentrated on the surface of a sphere of radius ct. This is its *support.*

3. It is a Retarded Green's Function in the following sense. When it is convolved with a source function $S(\boldsymbol{x}', t')$ to form a field quantity F

   $$F(\boldsymbol{x}, t) = \int\int\int\int G_R^{(3,1)}(\boldsymbol{x} - \boldsymbol{x}', t - t')S(\boldsymbol{x}', t')d\boldsymbol{x}'dt'$$

   we can evaluate the t' integration to yield

   $$F(\boldsymbol{x}, t) = \int\int\int \frac{S(\boldsymbol{x}', t - |\boldsymbol{x} - \boldsymbol{x}'|/c)}{4\pi|\boldsymbol{x} - \boldsymbol{x}'|}d\boldsymbol{x}'$$

   which is an integral over the spatial extent of the source operating at retarded times, i.e. at all times preceding the signal measurement time and which accounts for the travel time between source and receiver.

The first method of evaluation of the Retarded Green's Function used the evaluation of the temporal Fourier integral and then the resulting spatial integrals in that order. The second method we now discuss is a direct spatial Fourier transform of (2.5.2) to a one-dimensional equation in time, and a connection to our one-dimensional results in previous sections. Since G is a function of only difference variables we write

$$(\nabla^2 - c^{-2}\partial_t^2) G_R^{(3,1)}(\boldsymbol{x},t) = -\delta(\boldsymbol{x})\delta(t) \ , \tag{2.5.37}$$

and assume the initial condition

$$G_R^{(3,1)}(\boldsymbol{x},t) = 0, \qquad t < 0 \ , \tag{2.5.38}$$

emphasizing that the Retarded Green's Function vanishes for negative times. Next introduce the spatial Fourier transform pair

$$G_R^{(3,1)}(\boldsymbol{x},t) = (2\pi)^{-3} \int\!\!\int\!\!\int e^{i\boldsymbol{k}\cdot\boldsymbol{x}}\, \tilde{G}_R^{(3,1)}(\boldsymbol{k},t) d\boldsymbol{k} \ , \tag{2.5.39}$$

and

$$\tilde{G}_R^{(3,1)}(\boldsymbol{k},t) = \int\!\!\int\!\!\int e^{-i\boldsymbol{k}\cdot\boldsymbol{x}}\, G_R^{(3,1)}(\boldsymbol{x},t) d\boldsymbol{x} \ . \tag{2.5.40}$$

Take the spatial Fourier transform of (2.5.37). The result is a one-dimensional equation in t, parametric in $\boldsymbol{k}$,

$$\left(\frac{d^2}{dt^2} + c^2k^2\right) \tilde{G}_R^{(3,1)}(\boldsymbol{k},t) = c^2\delta(t) \ , \tag{2.5.41}$$

whose solution can be written as

$$\tilde{G}_R^{(3,1)}(\boldsymbol{k},t) = \begin{cases} G^+(\boldsymbol{k},t) = A\exp(ikct) + B\exp(-ikct), & t > 0 \ ; \\ G^-(\boldsymbol{k},t) = 0, & t < 0 \ . \end{cases} \tag{2.5.42}$$

The constants A and B are determined from the continuity condition

$$G^+(\boldsymbol{k},0) - G^-(\boldsymbol{k},0) = 0 \ , \tag{2.5.43}$$

and the jump discontinuity of the first derivative

$$\frac{d}{dt}G^+(\boldsymbol{k},0) - \frac{d}{dt}G^-(\boldsymbol{k},0) = c^2 \ , \tag{2.5.44}$$

which follows by integrating (2.5.41) from $-\varepsilon$ to ε. The result is

$$A = -B = c/2ik \ , \tag{2.5.45}$$

so that

$$\tilde{G}_R^{(3,1)}(\boldsymbol{k},t) = (c/2ik)(e^{ikct} - e^{-ikct})\theta(t) \ . \tag{2.5.46}$$

From (2.5.39) we have

$$G_R^{(3,1)}(\boldsymbol{x},t) = \frac{c\theta(t)}{2i(2\pi)^3}\int\int\int e^{i\boldsymbol{k}\cdot\boldsymbol{x}}(e^{ikct} - e^{-ikct})k^{-1}d\boldsymbol{k} \ , \tag{2.5.47}$$

which is the same spatial integral as (2.5.26). The result (2.5.35) follows.

We can also reproduce the analysis for other combinations of α and β. For $\alpha = \beta = 1$ we get the *Advanced* (A) Green's Function (see Fig. 2.3c)

$$G_A^{(3,1)}(x) = \frac{\theta(-x_0)}{4\pi r}\delta(x_0 + r) \ , \tag{2.5.48}$$

or

$$G_A^{(3,1)}(\boldsymbol{x},t) = \frac{\theta(-t)}{4\pi r}\delta(t + r/c) \ . \tag{2.5.49}$$

It is referred to as an Advanced Green's Function because if used to represent the effect of an arbitrary source distribution it yields an integral over the source function at times ahead (advanced) of the measurement time. When we use this function to find the corresponding Green's function for the Helmholtz equation in Sect. 2.8 we show that it corresponds to an incoming wave.

Similarly for $\alpha = \beta = 0$ (see Fig.2.3d), we get the principal value (P) Green's function which is

$$\begin{aligned} G_P^{(3,1)}(x) &= \delta(x^2)/4\pi \\ &= \Big[\delta(x_0 - r) + \delta(x_0 + r)\Big]/8\pi r \ , \end{aligned} \tag{2.5.50}$$

or

$$G_P^{(3,1)}(\boldsymbol{x},t) = \Big[\delta(t - r/c) + \delta(t + r/c)\Big]\Big/8\pi r \ , \tag{2.5.51}$$

where we have dropped the explicit temporal step functions. This will be shown to yield standing wave solutions for the Helmholtz equation.

Finally we easily see that if we drop the step function in (2.5.34), (2.5.35), (2.5.48) and (2.5.49) that

$$G_P^{(3,1)}(x) = \frac{1}{2}\Big[G_R^{(3,1)}(x) + G_A^{(3,1)}(x)\Big] \ , \tag{2.5.52}$$

so that only two of the three Green's functions are independent.

2.6 Green's Functions for the Wave Equation in Two Dimensions

Here we compute the Green's functions in two spatial and one temporal dimension. We do this by identifying one spatial coordinate and integrating the (3, 1) Green's functions over the source coordinate in this dimension. We choose the z-direction as our direction of integration and write the radius r as

$$r = \left[P^2 + (z - z')^2\right]^{1/2} , \tag{2.6.1}$$

where

$$\boldsymbol{P} = \boldsymbol{\rho} - \boldsymbol{\rho}' , \qquad \boldsymbol{\rho} = (x, y) , \tag{2.6.2}$$

will be the remaining two-dimensional vector. The Green's functions we define all satisfy the equation ($\tau = x_0 - x_0'$)

$$(\partial_x^2 + \partial_y^2 - c^{-2}\partial_t^2)\, G^{(2,1)}(\boldsymbol{P}, \tau) = -\delta(\boldsymbol{P})\delta(\tau) . \tag{2.6.3}$$

We first define the retarded Green's function $G_R^{(2,1)}$ as the spatial integral over $G_R^{(3,1)}$ given by (2.5.34) where $r = |\boldsymbol{x} - \boldsymbol{x}'|$ and $\tau = x_0 - x_0'$. It is

$$G_R^{(2,1)}(\boldsymbol{P}, \tau) = \int_{-\infty}^{\infty} G_R^{(3,1)}(x, x')dz' , \tag{2.6.4}$$

$$= \frac{1}{4\pi} \int_{-\infty}^{\infty} \frac{\delta(r - \tau)}{r} dz' . \tag{2.6.5}$$

To evaluate this integral, define the transformation

$$u = z - z' \qquad du = -dz'$$

$$r^2 = P^2 + u^2$$

$$du/r = dr/u = dr(r^2 - P^2)^{-1/2} , \tag{2.6.6}$$

so that

$$G_R^{(2,1)}(\boldsymbol{P}, \tau) = \frac{1}{2\pi} \int_{-\infty}^{\infty} \frac{\delta(r - \tau)dr}{(r^2 - P^2)^{1/2}} ,$$

and, evaluating the δ-function, we get

$$G_R^{(2,1)}(\boldsymbol{P}, \tau) = \frac{1}{2\pi} \frac{\theta(\tau - P)}{(\tau^2 - P^2)^{1/2}} . \tag{2.6.7}$$

This illustrates the fact that, in two dimensions, the effect of an impulse after a (scaled) time τ has elapsed has spread over a region of spatial extent $P < \tau$. A line source in three dimensions produces a field which, at any given point, has a tail. The support of the function is no longer confined to the surface of a sphere as it was in three dimensions.

We can analogously define the other Green's functions as integrals over the corresponding $(3,1)$ dimensional Green's functions. From (2.5.48) we get the *Advanced Green's Function*

$$G_A^{(2,1)}(\boldsymbol{P},\tau) = \frac{1}{2\pi}\frac{\theta(-\tau - P)}{(\tau^2 - P^2)^{1/2}} \tag{2.6.8}$$

and from (2.5.50) the *Principal Value Green's Function*

$$G_P^{(2,1)}(\boldsymbol{P},\tau) = \frac{1}{4\pi}\frac{1}{(\tau^2 - P^2)^{1/2}}\Big[\theta(\tau - P) + \theta(-\tau - P)\Big] \ . \tag{2.6.9}$$

The condition (2.5.52) is fulfilled for these three functions also.

Finally we note that the differential (2.6.3) for all these functions follows by integrating (2.5.6) over z' and introducing the transformation $u = z - z'$. The result is (2.6.3) which is independent of z.

2.7 Green's Functions for the Wave Equation in One Dimension

We compute the Green's function in one spatial and one temporal dimension. We begin again by integrating over one of the remaining spatial dimensions. Define

$$\xi = x - x', \qquad \eta = y - y' \ , \tag{2.7.1}$$

so that

$$P = (\xi^2 + \eta^2)^{1/2} \ . \tag{2.7.2}$$

We single out the y-direction and integrate over y'.

We define the (1,1) retarded Green's function by integrating (2.6.7) over y'

$$G_R^{(1,1)}(\xi,\tau) = \int_{-\infty}^{\infty} G_R^{(2,1)}(\boldsymbol{P},\tau)dy' \tag{2.7.3}$$

$$= \frac{1}{2\pi}\int_{-\infty}^{\infty}\frac{\theta\left(\tau - (\xi^2+\eta^2)^{1/2}\right)}{\left(\tau^2 - \xi^2 - \eta^2\right)^{1/2}}d\eta \ . \tag{2.7.4}$$

Define $\gamma = (\tau^2 - \xi^2)^{1/2}$ and note that for $|\eta| > \gamma$ the integrand in (2.7.4) vanishes. It also vanishes if $|\xi| > \tau$. The result is

$$G_R^{(1,1)}(\xi,\tau) = \frac{\theta(\tau - |\xi|)}{2\pi} \int_{-\gamma}^{\gamma} \frac{d\eta}{(\gamma^2 - \eta^2)^{1/2}} \ . \tag{2.7.5}$$

By substituting $\eta = \gamma \sin\theta$ the integral in (2.7.5) is just π. Thus,

$$G_R^{(1,1)}(\xi,\tau) = \frac{1}{2}\theta(\tau - |\xi|) \ . \tag{2.7.6}$$

Similarly, using (2.6.8) and (2.6.9) and the above procedure, the *Advanced* and *Principal Value Green's Functions* are respectively

$$G_A^{(1,1)}(\xi,\tau) = \frac{1}{2}\theta(-\tau - |\xi|) \ , \tag{2.7.7}$$

and

$$G_P^{(1,1)}(\xi,\tau) = \frac{1}{4}\Big[\theta(\tau - |\xi|) + \theta(-\tau - |\xi|)\Big] \ . \tag{2.7.8}$$

Finally, by integrating (2.6.3) over y' as in (2.7.3), we derive the equation which each Green's function satisfies. It is

$$\left(\partial_x^2 - c^{-2}\partial_t^2\right) G^{(1,1)}(\xi,\tau) = -\delta(\xi)\delta(\tau) \ . \tag{2.7.9}$$

2.8 Green's Functions for the Helmholtz Equation

2.8.1 Three Dimensions

We can define the Green's functions for the Helmholtz equation in three spatial dimensions as the temporal Fourier transform of the Green's functions for the wave equation derived in Sect. 2.5. Using the four-dimensional formulation we define

$$G^{(3)}(\boldsymbol{x},\boldsymbol{x}',k_0) = \int_{-\infty}^{\infty} G^{(3,1)}(x,x') \exp(ik_0\tau) d\tau \ , \tag{2.8.1}$$

where $\tau = x_0 - x_0'$. They satisfy the Helmholtz equation given by the same Fourier transform operating on (2.5.6). It is

$$\left(\nabla_x^2 + k_0^2\right) G^{(3)}(\boldsymbol{x},\boldsymbol{x}') = -\delta(\boldsymbol{x} - \boldsymbol{x}') \ , \tag{2.8.2}$$

where $k_0 = \omega/c$ and ω is circular frequency. For the *Retarded Green's Function* we have from (2.5.34) that

$$G_R^{(3,1)}(x, x') = G_R^{(3,1)}(x - x') = (4\pi r)^{-1}\delta(\tau - r) \ . \tag{2.8.3}$$

Substitute this in (2.8.1) to get

$$G_R^{(3)}(\boldsymbol{x}, \boldsymbol{x}', k_0) = (4\pi r)^{-1}\exp(ik_0 r) \ , \tag{2.8.4}$$

which is an outgoing spherical wave, i.e. for harmonic time dependence

$$\exp(-i\omega t) = \exp(-ik_0 x_0) \ , \tag{2.8.5}$$

the wave travels in a positive radial direction and satisfies an outgoing radiation condition of the form

$$\lim_{r\to\infty}\left[\frac{\partial}{\partial r} - ik_0\right] G_R^{(3)} = 0(r^{-2}) \ . \tag{2.8.6}$$

It expresses the field at the receiver point $\boldsymbol{x}$ due to a point source located at $\boldsymbol{x}'$ in a homogeneous medium with sound speed c.

From (2.5.48) we have for the *Advanced Green's Function*

$$G_A^{(3,1)}(x, x') = G_A^{(3,1)}(x - x') = (4\pi r)^{-1}\delta(\tau + r) \ , \tag{2.8.7}$$

which when substituted into (2.8.1) yields

$$G_A^{(3)}(\boldsymbol{x}, \boldsymbol{x}', k_0) = (4\pi r)^{-1}\exp(-ik_0 r) \ . \tag{2.8.8}$$

For harmonic time dependence this is an incoming radial wave satisfying the radiation condition

$$\lim_{r\to\infty}\left[\frac{\partial}{\partial r} + ik_0\right] G_A^{(3)} = 0(r^{-2}) \ . \tag{2.8.9}$$

The *Principal Value Green's Function* can be found by substituting (2.5.50) in (2.8.1) or by noting that it is just half the sum of (2.8.4) and (2.8.8). The result is

$$G_P^{(3)}(\boldsymbol{x}, \boldsymbol{x}', k_0) = (4\pi r)^{-1}\cos(k_0 r) \ , \tag{2.8.10}$$

which for harmonic time dependence represents a standing wave. Note also that in contrast to the retarded and advanced functions, the Principal Value Green's Function is real.

2.8.2 Two Dimensions

The two-dimensional Helmholtz Green's functions are the temporal Fourier transforms of the Green's functions for the two-dimensional wave equation in Sect. 2.6. They are defined as

$$G^{(2)}(\boldsymbol{\rho}, \boldsymbol{\rho}', k_0) = \int_{-\infty}^{\infty} G^{(2,1)}(\boldsymbol{P}, \tau) \exp(ik_0\tau) d\tau \ , \tag{2.8.11}$$

and satisfy the Helmholtz equation given by the corresponding transform of (2.6.3) which becomes

$$\left(\partial_x^2 + \partial_y^2 + k_0^2\right) G^{(2)}(\boldsymbol{\rho}, \boldsymbol{\rho}', k_0) = -\delta(x - x')\delta(y - y') \ . \tag{2.8.12}$$

For the *Retarded Green's Function* we substitute (2.6.7) in (2.8.11) to get

$$G_R^{(2)}(\boldsymbol{\rho}, \boldsymbol{\rho}', k_0) = \frac{1}{2\pi} \int_P^{\infty} \frac{\exp(ik_0\tau)}{(\tau^2 - p^2)^{1/2}} \ . \tag{2.8.13}$$

If we make the subsitution

$$\tau = P \cosh \psi; \quad d\tau = P \sinh \psi d\psi \ ,$$

the integral becomes

$$G_R^{(2)}(\boldsymbol{\rho}, \boldsymbol{\rho}', k_0) = \frac{1}{4\pi} \int_{-\infty}^{\infty} \exp(ik_0 P \cosh \psi) d\psi \ . \tag{2.8.14}$$

Written from $-\infty$ to ∞ since the integrand is an even function of ψ. The integral (2.8.14) is a representation of the Hankel function

$$G_R^{(2)}(\boldsymbol{\rho}, \boldsymbol{\rho}', k_0) = \frac{i}{4} H_0^{(1)}(k_0 P) \ , \tag{2.8.15}$$

which is an outgoing cylindrical wave, i.e. asymptotically

$$H_0^{(1)}(k_0 P) \sim (2/\pi i k_0 P)^{1/2} \exp(ik_0 P) \ , \tag{2.8.16}$$

which spreads like a cylindrical wave with amplitude factor $P^{-1/2}$.

For the *Advanced Green's Function* substitute (2.6.8) into (2.8.11). The integration can be performed as above with an additional sign change in τ. The result is the incoming Hankel function

$$G_A^{(2)}(\boldsymbol{\rho}, \boldsymbol{\rho}', k_0) = -\frac{i}{4} H_0^{(2)}(k_0 P) \ , \tag{2.8.17}$$

which represents a cylindrical wave propagating in the direction of decreasing r.

The *Principal Value Green's Function* can be computed either directly by substituting (2.6.9) in (2.8.11) or as half the sum of (2.8.15) and (2.8.17). The result is

$$G_P^{(2)}(\boldsymbol{\rho}, \boldsymbol{\rho}') = -\frac{1}{4} N_0(k_0 P) \ , \tag{2.8.18}$$

where N_0 is the Neumann function.

2.8.3 One Dimension

We define the one-dimensional Helmholtz Green's function as the temporal Fourier transform of the one-dimensional Green's functions for the wave equation in Sect. 2.7. It is

$$G^{(1)}(x, x', k_0) = \int_{-\infty}^{\infty} G^{(1,1)}(\xi, \tau) \exp(ik_0\tau) d\tau \ , \tag{2.8.19}$$

where $\xi = x - x'$. Fourier transformation of the differential (2.7.9) yields the ordinary differential equation

$$\left[\frac{d^2}{dx^2} + k_0^2\right] G^{(1)}(x, x', k_0) = -\delta(x - x') \ , \tag{2.8.20}$$

which is the one-dimensional version of the Helmholtz equation satisfied by all the one-dimensional Green's functions below.

For the *Retarded Green's Function* substitute (2.7.6) in (2.8.19). Evaluation of the step function yields

$$G_R^{(1)}(x, x', k_0) = \frac{1}{2} \int_{|\xi|}^{\infty} \exp(ik_0\tau) d\tau \ . \tag{2.8.21}$$

Recall that this Green's function for the wave equation was computed with k_0 shifted to $k_{0+} = k_0 + i\varepsilon$. We can then directly evaluate the integral in (2.8.21) since the contribution at ∞ vanishes. The result is

$$G_R^{(1)}(x, x', k_0) = -(2ik_0)^{-1} \exp(ik_0|x - x'|) \ , \tag{2.8.22}$$

which is a one-dimensional wave which travels to the right. This was just the free-space Green's function computed in (2.3.13) (with slightly different notation).

For the *Advanced Green's Function* substitute (2.7.7) in (2.8.19). The integral evaluation proceeds in the same manner as the previous example except that here we note that the advanced Green's function is computed with k_0 shifted to $k_{0-} = k_0 - i\varepsilon$. The result is

$$G_A^{(1)}(x, x', k_0) = (2ik_0)^{-1} \exp(-ik_0|x - x'|) \ , \tag{2.8.23}$$

which for harmonic time dependence is a one-dimensional wave travelling to the left. It is the complex conjugate of (2.8.22).

The *Principal Value Green's Function* can be computed either directly from (2.7.8) or by combining half the sum of (2.8.22) and (2.8.23) to give

$$G_P^{(1)}(x, x', k_0) = -(2k_0)^{-1} \sin(k_0|x - x'|) \ , \tag{2.8.24}$$

which represents a standing wave and is a real function.

2.9 Integral Representations: Helmholtz Green's Functions

There are several useful integral representations of the Green's function for the Helmholtz equation. In three dimensions the latter is a spherical wave and the representations amount to expanding a spherical wave into either plane waves (spectral integrals) or cylindrical waves (Sommerfeld and Weyrich representations). In two dimensions we expand the resulting cylindrical wave into plane waves.

2.9.1 Weyl Representation

The first representation is an expansion of a spherical wave into plane waves. The Helmholtz Green's function in three dimensions satisfies the differential equation

$$\left(\nabla_x^2 + k_0^2\right) G^{(3)}(\boldsymbol{x}, \boldsymbol{x}', k_0) = -\delta(\boldsymbol{x} - \boldsymbol{x}') \ . \tag{2.9.1}$$

Fourier transform this equation with respect to $\boldsymbol{x}$, i.e. multiply the equation by

$$\int\!\!\int\!\!\int \exp\left[-i(k_x x + k_y y + k_z z)\right] dx\, dy\, dz \ , \tag{2.9.2}$$

to get the mixed representation

$$\tilde{G}^{(3)}(\boldsymbol{k}, \boldsymbol{x}', k_0) = \exp(-i\boldsymbol{k} \cdot \boldsymbol{x}')(k^2 - k_0^2)^{-1} \ , \tag{2.9.3}$$

where

$$k^2 = |\boldsymbol{k}|^2 = k_x^2 + k_y^2 + k_z^2 = k_t^2 + k_z^2 \ , \tag{2.9.4}$$

and $\boldsymbol{k}$ is the Fourier transform variable. Note that in (2.9.4), by defining the transverse part of the wavenumber

$$k_t = (k_x^2 + k_y^2)^{1/2} \ , \tag{2.9.5}$$

we have essentially picked out the z-direction as special. Of course, we could do this with any of the three directions. Using (2.9.3) the inverse Fourier transform is, thus,

$$G^{(3)}(\boldsymbol{x}, \boldsymbol{x}', k_0) = \frac{1}{(2\pi)^3} \int\int\int$$

$$\cdot \frac{\exp[i\,(k_x(x-x') + k_y(y-y') + k_z(z-z'))]}{(k_z - K)(k_z + K)} dk_x\, dk_y\, dk_z \ , \tag{2.9.6}$$

where

$$K = (k_0^2 - k_t^2)^{1/2} \ , \tag{2.9.7}$$

and where we have distinguished the poles in the integrand of (2.9.6) as poles in k_z. We evaluate (2.9.6) cylindrically, i.e., we do the k_z integral first using complex variables. We shift the poles by adding a small positive imaginary part to k_0 and hence to K. That is

$$k_0 \to k_0 + i\varepsilon \doteq k_{0+} \ ; \quad K \to K + i\varepsilon \doteq K_+ \ .$$

Then in the complex k_z-plane we have the result illustrated in Fig.2.4.

We evaluate the k_z-integral in (2.9.6) by closing the contour in the upper half plane ($z - z' > 0$) or in the lower half plane ($z - z' < 0$). The result is

$$\int \frac{\exp\Big[ik_z(z-z')\Big]}{(k_z - K_+)(k_z + K_+)} dk_z = \frac{\pi i}{K_+} \exp\Big[iK_+|z-z'|\Big] \ . \tag{2.9.8}$$

Using this in (2.9.6) and writing the remaining two integrals using the two-dimensional vectors

$$\boldsymbol{k}_t = (k_x, k_y) \ ; \qquad \boldsymbol{x}_t = (x, y) \ , \tag{2.9.9}$$

we get

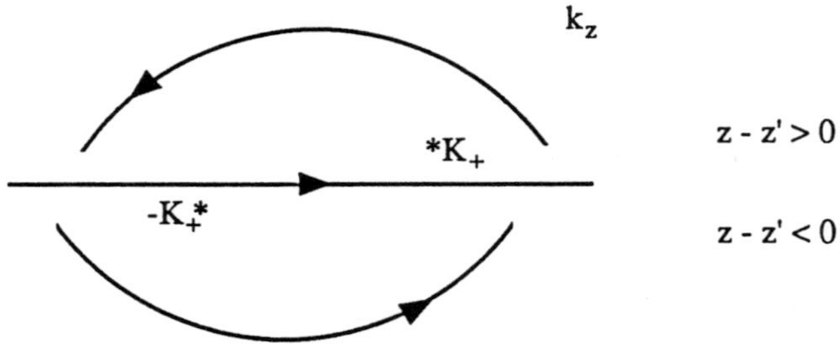

Fig. 2.4. Shifted integrand poles in the complex k_z-plane. The contour is closed with a semicircle whose radius approaches ∞ in the upper (lower) half plane for $z - z' > 0 (< 0)$.

$$G^{(3)}(\boldsymbol{x}, \boldsymbol{x}', k_0) = \frac{\pi i}{(2\pi)^3} \int\int \frac{\exp\left[i\boldsymbol{k}_t \cdot (\boldsymbol{x}_t - \boldsymbol{x}_t') + iK_+|z - z'|\right]}{K_+} d\boldsymbol{k}_t \, , \qquad (2.9.10)$$

where $\operatorname{Im} K_+ > 0$. This is just the Retarded Green's Function for the Helmholtz equation. We use this in the remainder of this section and, for convenience, drop the subscript R.

Equation (2.9.10) is the two-dimensional plane-wave spectral representation or *Weyl representation*. In deriving it we have singled out the z-direction as special. This is appropriate if the $z-$direction in the application is special, for example if there is a discontinuity in z or if the variability in the medium is in the z-direction. We treat this further in the next section. Also notice that here we shifted *both* poles by $+i\varepsilon$, so that effectively one shifted above the axis and one below. For the Retarded Green's Function in Sect. 2.5 we shifted both W poles down. This is equivalent to what we have done here since we had

$$k_0 - W \to k_0 - (W - i\varepsilon) = k_0 + i\varepsilon - W \, ,$$

$$k_0 + W \to k_0 + (W + i\varepsilon) = k_0 + i\varepsilon + W \, .$$

In both terms we give a positive shift to the k_0 term. By shifting the poles in the manner above we have derived the Weyl representation for the retarded Green's function. Other pole shifts can be done to form a Weyl representation for the advanced or principal value Green's function for example. We note the absolute value which occurs in (2.9.10). Care must be taken in using this representation in boundary integral representations, for example, if the boundary is not flat.

2.9.2 Sommerfeld Representation

For this case we expand the spherical wave in cylindrical waves. The expansion is essentially over the horizontal wave number. We begin by expressing the Weyl representation (2.9.10) in cylindrical polar coordinates defined about $x - x'$ and $y - y'$. We have that

$$\exp\left[i\boldsymbol{k}_t \cdot (\boldsymbol{x}_t - \boldsymbol{x}'_t)\right] = \exp(ik_t\rho\cos\theta) \ , \tag{2.9.11}$$

where

$$\rho = \left[(x-x')^2 + (y-y')^2\right]^{1/2} = |\boldsymbol{x}_t - \boldsymbol{x}'_t| \ , \tag{2.9.12}$$

and θ is the angle between $\boldsymbol{k}_t$ and $\boldsymbol{x}_t - \boldsymbol{x}'_t$. Using the cylindrical differential area element

$$d\boldsymbol{k}_t = k_t dk_t d\theta \ , \tag{2.9.13}$$

and the definition of the Bessel function (cylindrical wave)

$$J_0(k_t\rho) = (2\pi)^{-1} \int_0^{2\pi} \exp(ik_t\rho\cos\theta)d\theta \ , \tag{2.9.14}$$

we evaluate the θ-integral as above to get from (2.9.10)

$$G^{(3)}(\boldsymbol{x}, \boldsymbol{x}', k_0)$$

$$= \frac{i}{4\pi} \int_0^\infty \frac{J_0(k_t\rho)\exp\left[i\left(k_{0+}^2 - k_t^2\right)^{1/2}|z-z'|\right]}{(k_{0+}^2 - k_t^2)^{1/2}} k_t\, dk_t \ , \tag{2.9.15}$$

where we have explicitly written out the square root K. The result (2.9.15) is a one-dimensional integral representation of a spherical wave in terms of cylindrical waves called the *Sommerfeld representation.* It is an integral written over the *horizontal wavenumber* k_t, and is only useful for problems which contain an analogous horizontal symmetry (i.e. a parametric independence of θ).

Alternatively, we can write (2.9.15) in terms of the Hankel function. The Bessel function J_0 can be written in terms of Hankel functions $H_0^{(1)}$ and $H_0^{(2)}$ as

$$J_0(k_t\rho) = \frac{1}{2}\left[H_0^{(1)}(k_t\rho) + H_0^{(2)}(k_t\rho)\right] \ . \tag{2.9.16}$$

Substitute this into (2.9.15), use in the $H_0^{(2)}$ integral the result

$$H_0^{(2)}(k_t\rho) = -H_0^{(1)}(e^{\pi i}k_t\rho) \ , \tag{2.9.17}$$

and in this integral rotate the contour by defining a new variable

$$k_t' = e^{\pi i}k_t \ , \tag{2.9.18}$$

so that the limits of integration go from $(0, \infty)$ to $(0, \infty e^{\pi i}) = (0, -\infty)$. The result is an integral over only $H_0^{(1)}$ given by

$$G^{(3)}(\boldsymbol{x}, \boldsymbol{x}', k_0) = \frac{i}{8\pi}\int_{-\infty}^{\infty} \frac{H_0^{(1)}(k_t\rho)\exp\left[i\left(k_{0+}^2 - k_t^2\right)^{1/2}|z - z'|\right]}{(k_{0+}^2 - k_t^2)^{1/2}} k_t dk_t \ . \tag{2.9.19}$$

Since $H_0^{(1)}$ behaves asymptotically like an outgoing wave, the representation (2.9.19) is useful in problems that contain this type of geometry, e.g. the exterior problem of scattering from a bounded object. Equation (2.9.15) on the other hand is useful for an interior representation, i.e., one which contains standing waves rather than outgoing waves.

2.9.3 Explicit Evaluation of $G^{(3)}$

We mentioned that our representations were for the retarded Green's function. We can explicitly exhibit this by evaluating all the integrals in $G^{(3)}$. From (2.9.6) we have that

$$G^{(3)}(\boldsymbol{x}, \boldsymbol{x}') = \frac{1}{(2\pi)^3}\int\int\int \frac{\exp\left[i\boldsymbol{k}\cdot(\boldsymbol{x} - \boldsymbol{x}')\right]}{k^2 - k_0^2} d\boldsymbol{k} \ . \tag{2.9.20}$$

Using spherical polar coordinates defined as in Fig.2.5 we have that

$$\begin{aligned} \boldsymbol{k}\cdot(\boldsymbol{x} - \boldsymbol{x}') &= k|\boldsymbol{x} - \boldsymbol{x}'|\cos\theta = kr\cos\theta \ , \\ d\boldsymbol{k} &= k^2 dk \sin\theta d\theta d\phi \ . \end{aligned} \tag{2.9.21}$$

The ϕ-integral just yields 2π. The θ-integral is

$$\int_0^{\pi} \exp(ikr\cos\theta)\sin\theta d\theta = \Big[\exp(ikr) - \exp(-ikr)\Big](ikr)^{-1} \ . \tag{2.9.22}$$

We evaluate the positive exponential integral using complex variables and pole shifts to $k_{0+} = k_0 + i\varepsilon$ to get (close in *uhp*)

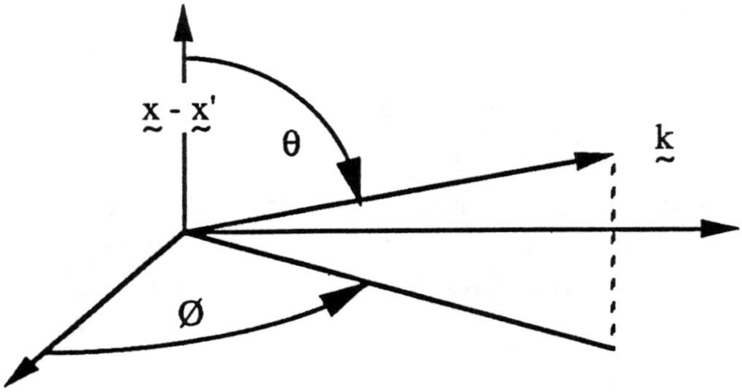

Fig. 2.5. Spherical polar coordinate system for the evaluation of (2.9.20).

$$\int_{-\infty}^{\infty} \frac{k \exp(ikr)}{(k - k_{0+})(k + k_{0+})} dk = \pi i \exp(ik_0 r) \ . \tag{2.9.23}$$

The negative exponential is evaluated by closing in the lower half plane with the *same* pole shifts $(k_0 + i\varepsilon)$ to yield

$$\int_{-\infty}^{\infty} \frac{k \exp(-ikr)}{(k - k_{0+})(k + k_{0+})} dk = -\pi i \exp(ik_0 r) \ . \tag{2.9.24}$$

Combining all these results we get

$$G^{(3)}(\boldsymbol{x},\, \boldsymbol{x}',\, k_0) = \exp(ik_0 r)/4\pi r \ , \tag{2.9.25}$$

which was the same result as we found using the retarded contour in Sect. 2.5.

2.9.4 Weyrich Representation

An alternative representation of spherical waves expanded in cylindrical waves can be found by expanding in the vertical (k_z) wavenumber rather than the horizontal wavenumber as in the Sommerfeld representation. Starting with (2.9.20) we don't do the k_z-integral. Instead do the k_x and k_y integrals using cylindrical symmetry. Using the definition

$$K_z = (k_0^2 - k_z^2)^{1/2} \ , \tag{2.9.26}$$

(2.9.20) is written as

$$G^{(3)}(\boldsymbol{x},\,\boldsymbol{x}',\,k_0) = \frac{1}{(2\pi)^3}\int_{-\infty}^{\infty} \exp\left[ik_z(z-z')\right] dk_z$$

$$\cdot \int\!\!\int \frac{\exp\left[i\boldsymbol{k}_t\cdot(\boldsymbol{x}_t - \boldsymbol{x}_{t'})\right]}{k_t^2 - K_z^2} dk_x\, dk_y \ . \tag{2.9.27}$$

The latter two integrals can be evaluated using the cylindrical coordinate and Bessel function definitions in (2.9.11) - (2.9.14). The θ-integral again yields a Bessel function so that we have

$$G^{(3)}(\boldsymbol{x},\,\boldsymbol{x},\,k_0) = \frac{1}{(2\pi)^2}\int_{-\infty}^{\infty} \exp\Big[ik_z(z-z')\Big] dk_z$$

$$\cdot \int_0^{\infty} \frac{k_t J_0(k_t\rho)}{k_t^2 - K_z^2} dk_t \ . \tag{2.9.28}$$

We use (2.9.16) for the Bessel function and in the integral for $H_0^{(2)}$ we rotate the contour ($k_t \to -k_t$) so that using (2.9.17) we again have an integral only over $H_0^{(1)}$. We get

$$G^{(3)}(\boldsymbol{x},\,\boldsymbol{x}',\,k_0) = \frac{1}{2(2\pi)^2}\int_{-\infty}^{\infty} \exp\Big[ik_z(z-z')\Big] dk_z$$

$$\cdot \int_{-\infty}^{\infty} \frac{k_t H_0^{(1)}(k_t\rho)}{k_t^2 - K_z^2} dk_t \ . \tag{2.9.29}$$

We have shifted our poles so that $\operatorname{Im} K_z > 0$. We evaluate the k_t-integral using complex variables by closing the contour in the *uhp*. Explicitly writing out K_z we get

$$G^{(3)}(\boldsymbol{x},\,\boldsymbol{x}',\,k_0) = \frac{i}{8\pi}\int_{-\infty}^{\infty} \exp\Big[ik_z(z-z')\Big] H_0^{(1)}\Big[\Big(k_{0+}^2 - k_z^2\Big)^{1/2}\rho\Big] dk_z \ , \tag{2.9.30}$$

which is *Weyrich's formula*, a one-dimensional integral representation in terms of the vertical wavenumber k_z. The result (2.9.30) is often quoted using (2.9.25) as

$$\frac{\exp\left[ik_0(\rho^2+z^2)^{1/2}\right]}{(\rho^2+z^2)^{1/2}} = \frac{i}{2}\int_{-\infty}^{\infty} e^{i\alpha z} H_0^{(1)}\left[\rho(k_0^2-\alpha^2)^{1/2}\right] d\alpha \ , \tag{2.9.31}$$

where ρ, z are real, $r = (\rho^2+z^2)^{1/2}$ and $0 \le \arg(k_0^2-\alpha^2)^{1/2} < \pi$. One advantage of this representation is that it does not contain an absolute value term.

2.9.5 Plane-Wave Decomposition of $G^{(2)}$

It is possible to derive two Weyl-type representations for the Helmholtz Green's function in two dimensions. In two dimensions the latter satisfies the Helmholtz equation

$$\left(\nabla_2^2 + k_0^2\right) G^{(2)}(\boldsymbol{x}_t, \boldsymbol{x}'_t, k_0) = -\delta(\boldsymbol{x}_t - \boldsymbol{x}'_t) \ . \tag{2.9.32}$$

We Fourier transform the equation by multiplying by

$$\int\int \exp\Big(-i\boldsymbol{k}_t \cdot \boldsymbol{x}_t\Big) d\boldsymbol{x}_t \ ,$$

where $\boldsymbol{k}_t = (k_x, k_y)$ is the Fourier transform variable. The result is the mixed representation

$$\tilde{G}^{(2)}(\boldsymbol{k}_t, \boldsymbol{x}'_t, k_0) = \exp(-i\boldsymbol{k}_t \cdot \boldsymbol{x}'_t)(k_t^2 - k_0^2)^{-1} \ . \tag{2.9.33}$$

The inverse transform thus becomes

$$G^{(2)}(\boldsymbol{x}_t, \boldsymbol{x}'_t, k_0) = \frac{1}{(2\pi)^2} \int\int \frac{\exp\left[i\boldsymbol{k}_t \cdot (\boldsymbol{x}_t - \boldsymbol{x}'_t)\right]}{(k_x - K_y)(k_x + K_y)} d\boldsymbol{k}_t \ , \tag{2.9.34}$$

where

$$K_y = (k_0^2 - k_y^2)^{1/2} \ . \tag{2.9.35}$$

We expressed the integrand of (2.9.34) in such a way as to do the k_x-integration using complex variable techniques just as we did the k_z-integration in (2.9.8). The result is

$$G^{(2)}(\boldsymbol{x}_t, \boldsymbol{x}'_t, k_0) = \frac{\pi i}{(2\pi)^2} \int_{-\infty}^{\infty} \frac{\exp\Big[ik_y(y - y') + iK_y|x - x'|\Big]}{K_y} dk_y \ . \tag{2.9.36}$$

Alternatively we could define

$$K_x = (k_0^2 - k_x^2)^{1/2} \tag{2.9.37}$$

so that the denominator of (2.9.34) becomes

$$(k_y - K_x)(k_y + K_x) \ ,$$

and the obvious choice is to carry out the k_y-integration. The result is

$$G^{(2)}(\boldsymbol{x}_t, \boldsymbol{x}'_t, k_0) = \frac{\pi i}{(2\pi)^2} \int_{-\infty}^{\infty} \frac{\exp\Big[ik_x(x - x') + iK_x|y - y'|\Big]}{K_x} dk_x \ . \tag{2.9.38}$$

The representations (2.9.36) and (2.9.38) are both Weyl-type representations for the two-dimensional Green's function. Their use depends on exploiting the geometry of the particular application.

2.9.6 Explicit Evaluation of $G^{(2)}$

We can explicitly evaluate $G^{(2)}$ using cylindrical coordinates and complex integration. We have from (2.9.34)

$$G^{(2)}(\boldsymbol{x}_t,\,\boldsymbol{x}'_t,\,k_0) = \frac{1}{(2\pi)^2}\int\!\!\int \frac{\exp\left[i\boldsymbol{k}_t\cdot(\boldsymbol{x}_t-\boldsymbol{x}'_t)\right]}{k_t^2-k_{0+}^2}dk_x\,dk_y \quad . \tag{2.9.39}$$

In cylindrical coordinates we have that

$$\begin{aligned} \boldsymbol{k}_t\cdot(\boldsymbol{x}_t-\boldsymbol{x}'_t) &= k_t|\boldsymbol{x}_t-\boldsymbol{x}'_t|\cos\theta = k_t\rho\cos\theta \ , \\ dk_x dk_y &= k_t dk_t d\theta \ . \end{aligned} \tag{2.9.40}$$

The θ-integration yields $2\pi J_0(k_t\rho)$, and we replace J_0 using (2.9.16). The integral involving $H_0^{(2)}$ is rewritten using (2.9.17) so that we only have an integral over $H_0^{(1)}$. It is

$$G^{(2)}(\boldsymbol{x}_t,\,\boldsymbol{x}'_t,\,k_0) = \frac{1}{4\pi}\int_{-\infty}^{\infty}\frac{H_0^{(1)}(k_t\rho)}{k_t^2-k_{0+}^2}k_t dk_t \quad . \tag{2.9.41}$$

Since $H_0^{(1)}$ behaves like an outgoing wave, we evaluate (2.9.41) using complex integration by closing the contour in the upper half plane. The poles are shifted by $k_0 \to k_0 + i\varepsilon$. The result is the cylindrical Green's function

$$G^{(2)}(\boldsymbol{x}_t,\,\boldsymbol{x}'_t,\,k_0) = (i/4)H_0^{(1)}(k_0\rho) \quad . \tag{2.9.42}$$

It is most useful for problems in cylindrical coordinates which have no angular dependence. The same result was derived in (2.8.15).

2.9.7 Explicit Relation Between $G^{(3)}$ and $G^{(2)}$

In three dimensions we have that

$$G^{(3)}(\boldsymbol{x},\,\boldsymbol{x}',\,k_0) = \frac{1}{(2\pi)^3}\int\!\!\int\!\!\int \frac{\exp\left[i\boldsymbol{k}\cdot(\boldsymbol{x}-\boldsymbol{x}')\right]}{k^2-k_{0+}^2}dk_x\,dk_y\,dk_z \quad . \tag{2.9.43}$$

Break up these three integrals into a k_z-integral and a two-dimensional transverse integral as

$$G^{(3)}(\boldsymbol{x},\, \boldsymbol{x}',\, k_0) = \frac{1}{2\pi}\int \exp\Big[ik_z(z-z')\Big]dk_z$$

$$\cdot \frac{1}{(2\pi)^2}\int\!\!\int \frac{\exp\Big[i\boldsymbol{k}_t\cdot(\boldsymbol{x}_t-\boldsymbol{x}'_t)\Big]}{k_t^2-\Big[k_0^2-k_z^2\Big]}dk_x\, dk_y \quad .(2.9.44)$$

The latter two integrals are just (2.9.39) but with k_0^2 replaced by $K_z^2 = k_0^2 - k_z^2$. We thus have that

$$G^{(3)}(\boldsymbol{x},\, \boldsymbol{x}',\, k_0) = \frac{1}{2\pi}\int_{-\infty}^{\infty} \exp\Big[ik_z(z-z')\Big]G^{(2)}(\boldsymbol{x}_t,\, \boldsymbol{x}'_t,\, K_z)dk_z \quad , \tag{2.9.45}$$

which explicitly relates $G^{(3)}$ to a value of $G^{(2)}$ via Fourier transformation.

2.10 Delta Functions in Different Coordinate Systems

The Green's function is characterized by satisfying a partial differential equation with a delta function source. In the previous sections of this chapter, we treated these delta functions in Cartesian coordinates. In future chapters, we need representations of this delta function in cylindrical and spherical coordinates. Here we describe a method of finding these representations which actually has more generality than just the application to cylindrical and spherical coordinates we describe.

2.10.1 Cylindrical Coordinates

We begin with the integral which defines the sifting property of the two dimensional Cartesian delta function

$$\int\!\!\int f(x,y)\delta(x-x_0)\delta(y-y_0)dxdy = f(x_0,y_0) \quad , \tag{2.10.1}$$

where f is an arbitrary test function. If we introduce the transformation

$$x = u(r,\theta) = r\cos\theta, \quad y = v(r,\theta) = r\sin\theta \quad , \tag{2.10.2}$$

whose Jacobian is

$$J = \begin{vmatrix} \partial u/\partial r & \partial u/\partial\theta \\ \partial v/\partial r & \partial v/\partial\theta \end{vmatrix} = r \quad ,$$

and differential element is

$$dx\,dy = |J|\,dr d\theta \quad , \tag{2.10.3}$$

we can rewrite (2.10.1) as

$$\int\int f(u,v)\delta(u-x_0)\delta(v-y_0)\,|J|\,dr\,d\theta = f(x_0,y_0) \quad . \tag{2.10.4}$$

This has the following interpretation: the distribution $\delta(u-x_0)\delta(v-y_0)|J|$ assigns to $f(u,v)$ the value $u = x_0$, $v = y_0$ or equivalently the value $r = r_0$, $\theta = \theta_0$ where we have $x_0 = u(r_0,\theta_0)$, $y_0 = v(r_0,\theta_0)$. This only works if the transformation is one-one, since we have to invert it to explicitly find r_0 and θ_0 as functions of x_0 and y_0. Further, the distribution $\delta(r-r_0)\delta(\theta-\theta_0)$ also assigns to $f(u,v)$ the value $r = r_0, \theta = \theta_0$, and in distribution space we have the equality

$$\delta(u-x_0)\delta(v-y_0)|J| = \delta(r-r_0)\delta(\theta-\theta_0) \quad , \tag{2.10.5}$$

or

$$\delta(x-x_0)\delta(y-y_0) = \frac{\delta(r-r_0)\delta(\theta-\theta_0)}{r} \quad . \tag{2.10.6}$$

Next, suppose the transformation is not one-one. That is, assume that J vanishes at some point. Here since $J = r$, the transformation is not one-one at $r = 0$ since θ can have any value (e.g. $r cos\theta = 0$ for any θ). Equivalently, we could suppose that our test function f is independent of θ, i.e. that θ is an *ignorable coordinate.* Then in (2.10.4) we can do the θ-integral since f and $|J|$ are both independent of θ. It becomes

$$\int f(u,v)\delta(u-x_0)\delta(v-y_0)\,|J|\,dr \int_0^{2\pi} d\theta = f(r_0) \quad , \tag{2.10.7}$$

where the *rhs* only depends on r_0. The θ-integral is just 2π. Then the distribution $\delta(u-x_0)\delta(v-y_0)\,|J|\,2\pi$ assigns to the test function $f(u,v)$ the value $u = x_0$, $v = y_0$ or equivalently $r = r_0$. So does the distribution $\delta(r-r_0)$ and these are thus equivalent

$$\delta(u-x_0)\delta(v-y_0)\,|J|\,2\pi = \delta(r-r_0) \quad , \tag{2.10.8}$$

or since $u = x$, $v = y$

$$\delta(x-x_0)\delta(y-y_0) = \frac{\delta(r-r_0)}{2\pi r} \quad . \tag{2.10.9}$$

This yields a representation of the two-dimensional Cartesian delta-function in terms of a cylindrical angle independent delta-function.

If we have three-dimensional cylindrical coordinates the additional $\delta(z-z_0)$ term is unchanged, i.e. the generalization of (2.10.6) is

$$\delta(x-x_0)\delta(y-y_0)\delta(z-z_0) = \frac{\delta(r-r_0)\delta(\theta-\theta_0)\delta(z-z_0)}{r} \; , \quad (2.10.10)$$

or, if independent of θ, the generalization of (2.10.9) is

$$\delta(x-x_0)\delta(y-y_0)\delta(z-z_0) = \frac{\delta(r-r_0)\delta(z-z_0)}{2\pi r} \; . \quad (2.10.11)$$

2.10.2 Spherical Coordinates

In three dimensions the sifting property is

$$\int\int\int f(x,y,z)\delta(x-x_0)\delta(y-y_0)\delta(z-z_0)dx\,dy\,dz = f(x_0,y_0,z_0) \; . \quad (2.10.12)$$

Introduce the transformation ($\theta:\ 0\to\pi$ and $\phi:\ 0\to 2\pi$)

$$\begin{aligned} x &= u(r,\theta,\phi) = r\sin\theta\cos\phi \\ y &= v(r,\theta,\phi) = r\sin\theta\sin\phi \\ z &= w(r,\theta,\phi) = r\cos\theta \; , \end{aligned} \quad (2.10.13)$$

with

$$J = \begin{vmatrix} \partial u/\partial r & \partial u/\partial\theta & \partial u/\partial\phi \\ \partial v/\partial r & \partial v/\partial\theta & \partial v/\partial\phi \\ \partial w/\partial r & \partial w/\partial\theta & \partial w/\partial\phi \end{vmatrix} = r^2\sin\theta \; , \quad (2.10.14)$$

and

$$dx\,dy\,dz = |J|\,dr\,d\theta\,d\phi \; , \quad (2.10.15)$$

so that (2.10.12) is

$$\int\int\int f(u,v,w)\delta(u-x_0)\delta(v-y_0)\delta(w-z_0)\,|J|\,dr\,d\theta\,d\phi = f(x_0,y_0,z_0) \; . \quad (2.10.16)$$

This has the interpretation: the distribution $\delta(u-x_0)\delta(v-y_0) \cdot\delta(w-z_0)\,|J|$ assigns to the test function $f(u,v,w)$ the value $u=x_0$, $v=y_0$, $w=z_0$, or equivalently the value r_0 , θ_0 , ϕ_0 found via inverting

$$\begin{aligned} x_0 &= u(r_0,\, \theta_0,\, \phi_0) \\ y_0 &= v(r_0,\, \theta_0,\, \phi_0) \\ z_0 &= w(r_0, \theta_0, \phi_0) \ . \end{aligned} \tag{2.10.17}$$

The transformation must be one-one since we have to invert it to find r_0, θ_0, ϕ_0 as functions of x_0, y_0, z_0. The distribution $\delta(r - r_0)\delta(\theta - \theta_0)\delta(\phi - \phi_0)$ has the same effect so that we have

$$\delta(u - x_0)\, \delta(v - y_0)\, \delta(w - z_0)\, |J|$$

$$= \delta(r - r_0)\, \delta(\theta - \theta_0)\, \delta(\phi - \phi_0) \ , \tag{2.10.18}$$

or, using (2.10.14) we write two representations

$$\delta(x - x_0)\, \delta(y - y_0)\, \delta(z - z_0)$$

$$= \frac{\delta(r - r_0)\, \delta(\theta - \theta_0)\, \delta(\phi - \phi_0)}{r^2 \sin\theta} \ , \tag{2.10.19}$$

$$= \frac{\delta(r - r_0)\, \delta(\cos\theta - \cos\theta_0)\, \delta(\phi - \phi_0)}{r^2} \ . \tag{2.10.20}$$

Note that the absolute value on $\sin\theta$ is dropped in (2.10.19) since in the range $\theta :\ 0 \to \pi$ it is positive.

Next, suppose the transformation is not one-one. J can vanish in either of two ways:

(a) $r = 0$, in which case both θ and ϕ are ignorable coordinates, and equivalently the test function f is independent of θ and ϕ.

(b) $\theta = 0$. These are points on the z-axis. Here only ϕ is ignorable since if $\theta = 0$ and $z \neq 0$ on the z-axis, then $r \neq 0$ and r is not ignorable. Equivalently we have that the test function f is independent of ϕ.

For (a), we can do the θ and ϕ integrals in (2.10.16). Here f is independent of both θ and ϕ but J depends on θ so we have

$$\int f(u, v, w)\, \delta(u - x_0)\, \delta(v - y_0)\delta(w - z_0) dr \int\!\!\int |J|\, d\theta d\phi = f(r_0) \ . \tag{2.10.21}$$

The angle integrals just equal $4\pi r^2$, so that we have that the distribution $\delta(u-x_0)\,\delta(v-y_0)\,\delta(w-z_0)4\pi r^2$ assigns to the test function $f(u,v,w)$ the value $u=x_0, v=y_0$, $w=z_0$ or equivalently the value $r=r_0$. So does the distribution $\delta(r-r_0)$ and we thus have the equality

$$\delta(u-x_0)\,\delta(v-y_0)\,\delta(w-z_0)4\pi r^2 = \delta(r-r_0) \ , \qquad (2.10.22)$$

or

$$\delta(x-x_0)\,\delta(y-y_0)\,\delta(z-z_0) = \frac{\delta(r-r_0)}{4\pi r^2} \ . \qquad (2.10.23)$$

This equates the three-dimensional Cartesian delta function to a one-dimensional radial delta function where we have independence of both angles.

In (b), do the ϕ integral in (2.10.16) so that it becomes

$$\int\!\!\int f(u,v,w)\,\delta(u-x_0)\,\delta(v-y_0)\delta(w-z_0)\,|J|\,dr\,d\theta \int d\phi = f(r_0,\theta_0) \ . \qquad (2.10.24)$$

The ϕ-integral is just 2π so that the distribution $\delta(u-x_0)\ \delta(v-y_0)$ $\cdot\,\delta(w-z_0)2\pi\,|J|$ assigns to the test function f the value $r=r_0$, $\theta=\theta_0$. So does the distribution $\delta(r-r_0)\delta(\theta-\theta_0)$ and we have the equality

$$\begin{aligned}\delta(x-x_0)\,\delta(y-y_0)\,\delta(z-z_0)& \\ &= \frac{\delta(r-r_0)\,\delta(\theta-\theta_0)}{2\pi r^2\sin\theta} \ , \\ &= \frac{\delta(r-r_0)\delta(\cos\theta-\cos\theta_0)}{2\pi r^2} \ . \qquad (2.10.25)\end{aligned}$$

2.11 Other Examples of Green's Functions

In this section we consider other examples of Green's functions for cylindrical and spherical coordinates. Specifically we solve for the two-dimensional Green's function in angle-independent cylindrical coordinates using two methods, the Hankel transform method in Example 1 and the Fourier transform method in Example 2. In Example 3 we solve for the one-dimensional Green's function which arises in angle-independent spherical coordinates.

Example 1: Two-Dimensional Green's Function Using Hankel Transforms

In cylindrical coordinates the angle-independent two-dimensional Green's function satisfies the differential equation

$$\frac{1}{r}\frac{\partial}{\partial r}\left(r\frac{\partial g}{\partial r}\right)+\frac{\partial^2 g}{\partial z^2}+k_0^2 g=\frac{-\delta(r-r_0)\,\delta(z-z_0)}{2\pi r}\ , \tag{2.11.1}$$

where $g = g(r, z; r_0, z_0)$ and where the source point (r_0, z_0) is arbitrary. Here $-\infty < z < \infty$ and $r > 0$. This equation can be solved in several ways. In this example we use Hankel transforms. In Example 2, we solve the problem using Fourier transforms.

Define the Hankel transform pair

$$g_H(p, z; r_0, z_0) = \int_0^\infty r J_0(pr) g(r, z; r_0, z_0) dr\ , \tag{2.11.2}$$

whose inverse is

$$g(r, z; r_0, z_0) = \int_0^\infty p\, J_0(pr) g_H(p, z; r_0, z_0) dp\ . \tag{2.11.3}$$

Next, Hankel transform (2.11.1), i.e. multiply it by

$$\int_0^\infty r J_0(pr) dr\ . \tag{2.11.4}$$

The result is

$$\int_0^\infty J_0(pr)\frac{\partial}{\partial r}\left(r\frac{\partial g}{\partial r}\right)dr+\frac{d^2}{dz^2}g_H+k_0^2 g_H$$

$$=\frac{-\delta(z-z_0)}{2\pi}\,J_0(pr_0)\ . \tag{2.11.5}$$

Integrate the first term in (2.11.5) by parts twice

$$\int_0^\infty J_0(pr)\frac{\partial}{\partial r}\left(r\frac{\partial g}{\partial r}\right)dr = r\frac{\partial g}{\partial r}J_0(pr)\Big|_0^\infty - rg\frac{\partial}{\partial r}J_0(pr)\Big|_0^\infty$$

$$+\int_0^\infty g\,\frac{\partial}{\partial r}\left[r\frac{\partial}{\partial r}J_0(pr)\right]dr\ . \tag{2.11.6}$$

We assume the terms integrated by parts all vanish (and check the assumption below) and use the differential equation for the Bessel function

$$\frac{\partial}{\partial r}\left[r\frac{\partial}{\partial r}J_0(pr)\right] = -rp^2 J_0(pr)\ . \tag{2.11.7}$$

Equation (2.11.5) can thus be written

$$\frac{d^2}{dz^2} g_H + K^2 g_H = \frac{-\delta(z - z_0)}{2\pi} J_0(pr_0) \ , \tag{2.11.8}$$

where $K^2 = k_0^2 - p^2$. This has the free-space solution

$$g_H(p, z; r_0, z_0) = \frac{e^{iK|z-z_0|}}{-2iK} \frac{J_0(pr_0)}{2\pi} \ , \tag{2.11.9}$$

and the corresponding inverse transform yields

$$g(r, z; r_0, z_0) = \frac{i}{4\pi} \int_0^\infty J_0(pr) J_0(pr_0) \frac{e^{iK|z-z_0|}}{K} pdp \ . \tag{2.11.10}$$

For the special case $r_0 = 0$, (2.11.10) is just Sommerfeld's formula

$$g(r, z; 0, z_0) = \frac{1}{4\pi} \frac{e^{ik_0 \left[r^2 + (z-z_0)^2\right]^{1/2}}}{\left[r^2 + (z - z_0)^2\right]^{1/2}} \ . \tag{2.11.11}$$

As a check we note that the terms we dropped using partial integration in (2.11.7) do indeed vanish. For $\operatorname{Im} k_0 > 0$, the infinite limits vanish, and the $r = 0$ limits vanish since g and $\partial g / \partial r$ are finite at $r = 0$ for $z \neq z_0$.

Finally we note that the integration variable p in (2.11.10) is just the radial part of the wavenumber. The integral in (2.11.10) is on p because we directly transformed and inverse transformed using the radial variable and its transform conjugate.

Example 2: Two-Dimensional Green's Function Using Fourier Transforms

An alternate solution of (2.11.1) in Example 1 is to first Fourier transform in depth. Define the Fourier transform

$$\tilde{g}(r, q; r_0, z_0) = \int_{-\infty}^{\infty} e^{-iq(z-z_0)} g(r, z; r_0, z_0) dz \ , \tag{2.11.12}$$

and apply the transform to (2.11.1). The result is

$$\frac{d}{dr}\left[r \frac{d}{dr} \tilde{g}\right] + r M^2 \tilde{g} = \frac{-\delta(r - r_0)}{2\pi} \tag{2.11.13}$$

where $M^2 = k_0^2 - q^2$. This is just the equation for the one-dimensional Green's function in cylindrical coordinates. The solutions of the homogeneous equation are Bessel functions. We write

$$\tilde{g}(r,q;r_0,z_0) = \begin{cases} \tilde{g}^+(r,q;r_0,z_0) = AH_0^{(1)}(Mr) & r > r_0 \ , \\ \tilde{g}^-(r,q;r_0,z_0) = BJ_0(Mr) & r < r_0 \ , \end{cases} \tag{2.11.14}$$

where $\tilde{g}^+$ is chosen to have outgoing wave behavior as $r \to \infty$, and $\tilde{g}^-$ to be finite at $r = 0$. Continuity of $\tilde{g}$ at $r = r_0$

$$\tilde{g}^+(r_0,q;r_0,z_0) = \tilde{g}^-(r_0,q;r_0,z_0) \ , \tag{2.11.15}$$

and the jump discontinuity of its derivative (from integrating (2.11.13))

$$\frac{d}{dr}\tilde{g}^+(r_0,q;r_0,z_0) - \frac{d}{dr}\tilde{g}^-(r_0,q;r_0,z_0) = -\frac{1}{2\pi r_0} \ , \tag{2.11.16}$$

are used to solve for the constants A and B. The result is

$$\tilde{g}(r,q;r_0,z_0) = \frac{i}{4}\begin{cases} J_0(Mr_0)H_0^{(1)}(Mr) & r > r_0 \\ J_0(Mr)H_0^{(1)}(Mr_0) & r < r_0 \end{cases} \tag{2.11.17}$$

and its Fourier inverse is defined by

$$g(r,z;r_0,z_0) = \frac{1}{2\pi}\int_{-\infty}^{\infty} e^{iq(z-z_0)}\tilde{g}(r,q;r_0,z_0)dq \ , \tag{2.11.18}$$

which is written as

$$g(r,z;r_0,z_0) = \frac{i}{8\pi}\int_{-\infty}^{\infty} e^{iq(z-z_0)}J_0(Mr_<)H_0^{(1)}(Mr_>)dq \ , \tag{2.11.19}$$

where $r_<$ and $r_>$ refer to the lesser or greater of (r, r_0). For $r_0 = 0$, this is just Weyrich's formula

$$\begin{aligned} g(r,z;0,z_0) &= \frac{i}{8\pi}\int_{-\infty}^{\infty} e^{iq(z-z_0)}H_0^{(1)}(Mr)dq \\ &= \frac{1}{4\pi}\frac{e^{ik_0\left[r^2+(z-z_0)\right]^{1/2}}}{\left[r^2+(z-z_0)^2\right]^{1/2}} \ . \end{aligned} \tag{2.11.20}$$

which is just (2.11.11).

Example 3: Spherical Coordinate

The inhomogeneous Helmholtz equation in one-dimension for an angle independent three-dimensional Green's function is given by

$$\frac{1}{r^2}\frac{d}{dr}\left[r^2\frac{dg}{dr}(r,r')\right]+k_0^2 g(r,r')=-\frac{\delta(r-r')}{4\pi r^2} \ . \qquad (2.11.21)$$

First, multiply through by r^2 to get

$$\frac{d}{dr}\left[r^2\frac{dg}{dr}(r,r')\right]+k_0^2\, r^2 g(r,r')=-\frac{\delta(r-r')}{4\pi} \ . \qquad (2.11.22)$$

The solutions of the homogeneous equation are

$$\frac{exp(\pm ik_0 r)}{r} \text{ or } \left(\frac{\cos k_0 r}{r}, \frac{\sin k_0 r}{r}\right) \ .$$

For $r > r'$ we choose the outgoing radial wave solution g^+, and for $r < r'$ that solution which is finite at $r = 0$, g^-. We write $g(r, r')$ as

$$g(r,r')=\begin{cases} A\ \dfrac{e^{ik_0 r}}{r} & r>r' \ , \\ B\ \dfrac{\sin(k_0 r)}{r} & r<r' \ . \end{cases} \qquad (2.11.23)$$

The function is continuous at $r = r'$

$$A\ \frac{e^{ik_0 r'}}{r'}=B\ \frac{\sin(k_0 r')}{r'} \ , \qquad (2.11.24)$$

and from the differential equation, has the jump discontinuity in its radial derivative given by

$$(r')^2\left[\frac{dg^+}{dr}(r',r')-\frac{dq^-}{dr}(r',r')\right]=-\frac{1}{4\pi} \ . \qquad (2.11.25)$$

Using (2.11.23) this becomes

$$A(ik_0 r'-1)e^{ik_0 r'}-B\Big[k_0 r'\cos(k_0 r')-\sin k_0 r'\Big]=-\frac{1}{4\pi} \ . \qquad (2.11.26)$$

We have from (2.11.24) that

$$A = B \sin(k_0 r') e^{-ik_0 r'} \quad , \tag{2.11.27}$$

so that (2.11.26) yields

$$B = \frac{e^{ik_0 r'}}{4\pi k_0 r'} \quad , \tag{2.11.28}$$

and thus

$$A = \frac{\sin k_0 r'}{4\pi k_0 r'} \quad , \tag{2.11.29}$$

so that we have

$$g(r, r') = \frac{1}{4\pi k_0 r r'} \begin{cases} \sin(k_0 r') e^{ik_0 r} \ , & r > r' \ , \\ \sin(k_0 r) e^{ik_0 r'} \ , & r < r \ . \end{cases} \tag{2.11.30}$$

In particular in the limit as $r' \to 0$

$$g(r, 0) = \frac{e^{ik_0 r}}{4\pi r} \quad , \tag{2.11.31}$$

which is the standard spherical wave solution for the source point located at $r' = 0$. For $r' \neq 0$ we have the representation (2.11.30).

2. Appendix

2.A Dirac-Plemelj Relations

In this appendix, we drive the Dirac-Plemelj relations used in the chapter. We consider integrals of the following form

$$F(z) = \int_a^b \frac{f(x')}{x' - z} dx' \quad , \tag{2.A.1}$$

where z is a complex variable and the integration variable x' is confined to the real axis (see Fig. 2.A(a)). These are similar to Cauchy integrals but without a closed path, and although we consider them in Cartesian coordinates, they can be generalized to any arc. The question we want to answer is the following: let z be on or near the real axis ($z \approx x$) in the region where the integral is defined ($a \leq x \leq b$). What meaning do we attach to the representation

$$F(x) = \int_a^b \frac{f(x')}{x' - x} dx' \quad ? \tag{2.A.2}$$

The meaning is clearly ambiguous since we have not specified how we take the limit. There are three possibilities ($\varepsilon > 0$)

$$\text{(a)} \quad F_+(x) = \lim_{\varepsilon \to 0} F(x + i\varepsilon) = \lim_{\varepsilon \to 0} \int_a^b \frac{f(x')}{x' - x - i\varepsilon} dx \quad , \tag{2.A.3}$$

$$\text{(b)} \quad F_-(x) = \lim_{\varepsilon \to 0} F(x - i\varepsilon) = \lim_{\varepsilon \to 0} \int_a^b \frac{f(x')}{x' - x + i\varepsilon} dx \quad , \tag{2.A.4}$$

and

(c) Cauchy Principal Value

$$P \int_a^b \frac{f(x')}{x' - x} dx' = \lim_{\varepsilon \to 0} \left\{ \int_a^{x-\varepsilon} \frac{f(x')}{x' - x} dx' + \int_{x+\varepsilon}^b \frac{f(x')}{x' - x} dx' \right\} \quad . \tag{2.A.5}$$

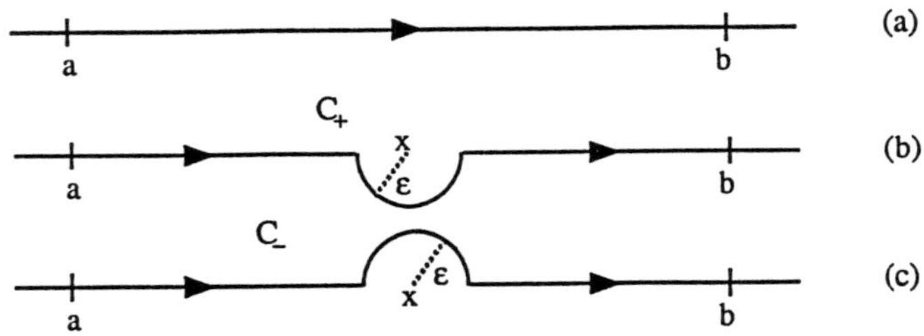

Fig. 2.A.1. (a) Real contour of integration, (b) Distorted contour C_+ for the integration of F_+, and (c) Distorted contour C_- for the integration of F_-.

The limits can be thought of as approaching the point x from above (+), below (−), and symmetricaly along the real axis.

We require that $f(x')$ have certain smoothness properties so that the contributions from opposite sides of the singular point will cancel. We thus assume that $f(z')$ is analytic near the real axis. We can thus use the Cauchy theorem to deform the contours in F_+ and F_-. We illustrate this with F_+. In (2.A.3) we shift the ε-term to x' and analytically continue so that we can write

$$F_+(x) = \int_{C_+} \frac{f(z')}{z' - x} dz' \ , \tag{2.A.6}$$

where the contour C_+ is illustrated in Fig. 2.A(b). This can be written as a sum of three integrals

$$F_+(x) = \lim_{\varepsilon \to 0} \left\{ \int_a^{x-\varepsilon} \frac{f(x')}{x' - x}\, dx' + \int_{x+\varepsilon}^{b} \frac{f(x')}{x' - x}\, dx' + \int_{\smile} \right\} \ , \tag{2.A.7}$$

the latter of which is over the lower semicircle defined by $z' = x + \varepsilon \exp(i\theta)$ where $\theta: \ -\pi$ to 0. The first two integrals form the Cauchy Principal Value from (2.A.5), and (2.A.7) becomes

$$F_+(x) = P \int_a^b \frac{f(x')}{x' - x} dx' + \lim_{\varepsilon \to 0} \int_{-\pi}^{0} \frac{\varepsilon e^{i\theta} f(x + \varepsilon e^{i\theta})}{\varepsilon e^{i\theta}}\, i d\theta \ , \tag{2.A.8}$$

or

$$F_+(x) = P \int_a^b \frac{f(x')}{x' - x} dx' + \pi i f(x) \ . \tag{2.A.9}$$

Comparison of (2.A.9) with the *rhs* of (2.A.3) yields the operator relationship

$$\frac{1}{x' - x - i\varepsilon} = P\frac{1}{x' - x} + \pi i\,\delta(x' - x)\ , \tag{2.A.10}$$

where the limit $\varepsilon \to 0$ is implicit. An analogous relation can be derived using F_- and a semicircular contour above the real axis (see Fig. 2.A(c)) to yield

$$\frac{1}{x' - x + i\varepsilon} = P\frac{1}{x' - x} - \pi i\delta(x' - x)\ . \tag{2.A.11}$$

The relations (2.A.10) and (2.A.11) are referred to as the Dirac-Plemelj relations.

3. Scalar Plane Wave Scattering

In this chapter we consider one of the simplest non-trivial scattering problems, that of plane wave scattering from an interface. The latter separates two different fluids (or two gases, or a fluid and a gas). The problem is highly idealized since both the incident plane wave and the interface have infinite extent. Nevertheless, we will learn some very general principles from this idealization which will follow us to more complicated and realistic problems. For most of the chapter the interface is considered to be flat, although in Sect. 3.8 we briefly consider the effect of roughness.

3.1 Scattering from a Plane Interface

We describe the scattering using a *velocity potential* ψ. The velocity is $-\partial_j\psi$ and, for time-harmonic waves, the pressure is $p = -i\omega\rho\psi$. This was discussed in Chap. 1. The interface at $z = 0$ separates two different media with constant densities ρ_j and sound speeds c_j (wavenumbers k_j) where $j = 1, 2$. In Fig. 3.1 we have schematically illustrated a plane wave incident on this interface at angle θ_i, a reflected field at angle θ_r and a transmitted field at angle θ_t.

In region 1 the velocity potential ψ_1 satisfies the differential equation

$$(\nabla^2 + k_1^2)\psi_1(x, z) = 0 \quad . \tag{3.1.1}$$

It is composed of incident (ψ_i) plus reflected (ψ_r) fields,

$$\psi_1(x, z) = \psi_i(x, z) + \psi_r(x, z) \quad , \tag{3.1.2}$$

where the incident and reflected plane wave fields are given by

$$\psi_i(x, z) = A_i \exp\Big\{ik_1(x\sin\theta_i - z\cos\theta_i)\Big\} \quad , \tag{3.1.3}$$

and

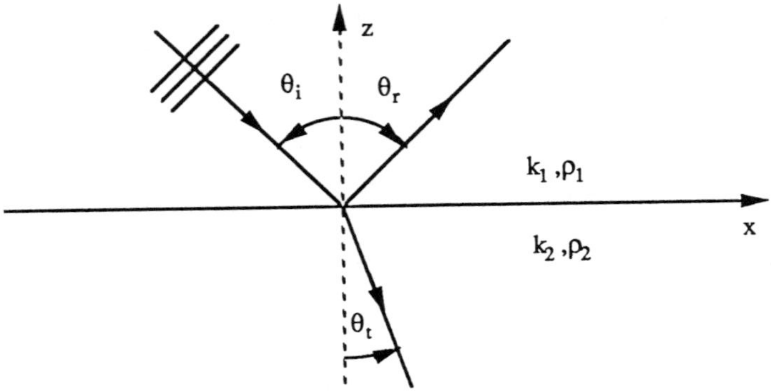

Fig. 3.1. Plane wave incident at angle θ_i on an interface separating media (1 and 2) with different wave speeds and densities.

$$\psi_r(x,z) = A_r \exp\Big\{ik_1(x\sin\theta_r + z\cos\theta_r)\Big\} \ . \tag{3.1.4}$$

We assume the amplitude A_i and direction θ_i of the incident field are known, and we must find the amplitude and direction of the reflected wave.

In region 2, ψ_2 satisfies the equation

$$\left(\nabla^2 + k_2^2\right)\psi_2(x,z) = 0 \ , \tag{3.1.5}$$

and it is equal to the transmitted field ψ_t

$$\psi_2(x,z) = \psi_t(x,z) = A_t \exp\Big\{ik_2(x\sin\theta_t - z\cos\theta_t)\Big\} \ , \tag{3.1.6}$$

whose amplitude and direction must be found. We have assumed it to be a plane wave traveling in the positive x-direction and negative z-direction.

Since the interface and the incident plane wave are all of infinite extent, the scattering problem should be independent of the particular x-point on the surface where the wave hits. This is another way of indicating that the problem is translationally invariant in x. This can only be true if the x-components of the phases in each of our waves are equal, i.e.

$$k_1\sin\theta_i = k_1\sin\theta_r = k_2\sin\theta_t \ . \tag{3.1.7}$$

From this we can immediately conclude that the angle of reflection equals the angle of incidence

$$\theta_r = \theta_i \ , \tag{3.1.8}$$

and Snell's Law

$$k_1 \sin\theta_i = k_2 \sin\theta_t \ , \tag{3.1.9}$$

which determines the angle of transmission from the incidence angle and the wave number ratio. Since $k_j = \omega/c_j$ where c_j is the sound speed in the medium, (3.1.9) can also be written as

$$c_2 \sin\theta_i = c_1 \sin\theta_t \ . \tag{3.1.10}$$

At the $z = 0$ interface, the pressure and normal component of velocity are continuous. These conditions can be written as

$$\rho_1\psi_1(x,0) = \rho_2\psi_2(x,0) \ , \tag{3.1.11}$$

and

$$\frac{\partial\psi_1}{\partial z}(x,0) = \frac{\partial\psi_2}{\partial z}(x,0) \ . \tag{3.1.12}$$

Using (3.1.2) - (3.1.4) and (3.1.6), (3.1.11) and (3.1.12) become

$$\rho_1 (A_i + A_r) = \rho_2 A_t \ , \tag{3.1.13}$$

and

$$k_1 \cos\theta_i (A_r - A_i) = -k_2 \cos\theta_t A_t \ , \tag{3.1.14}$$

which can be solved for the ratios $R = A_r/A_i$ and $T = A_t/A_i$ to yield

Version 1:

$$R = \frac{\rho_2 k_1 cos\theta_i - \rho_1 k_2 \cos\theta_t}{\rho_2 k_1 \cos\theta_i + \rho_1 k_2 \cos\theta_t} \ , \tag{3.1.15}$$

and

$$T = \frac{2\rho_1 k_1 \cos\theta_i}{\rho_2 k_1 \cos\theta_i + \rho_1 k_2 \cos\theta_t} \ . \tag{3.1.16}$$

These are the reflection and transmission coefficients, respectively. Both are scaled to the amplitude of the incident field. There are several other versions of R and T which are convenient to have available. Define the ratios:

$$\rho = \rho_2/\rho_1 \ ; \quad K = k_2/k_1 \tag{3.1.17}$$

and note that using Snell's law we can write

$$
\begin{aligned}
\cos\theta_t &= (1-\sin^2\theta_t)^{1/2} \\
&= \left(1-(k_1/k_2)^2\sin^2\theta_i\right)^{1/2} , \\
&= K^{-1}\left(K^2-\sin^2\theta_i\right)^{1/2} .
\end{aligned}
\tag{3.1.18}
$$

We can then eliminate the transmitted angle in Version 1 above and write

Version 2:

$$
R=\frac{\rho\cos\theta_i-(K^2-\sin^2\theta_i)^{1/2}}{\rho\cos\theta_i+(K^2-\sin^2\theta_i)^{1/2}} , \tag{3.1.19}
$$

and

$$
T=\frac{2\cos\theta_i}{\rho\cos\theta_i+(K^2-\sin^2\theta_i)^{1/2}} . \tag{3.1.20}
$$

Alternatively we could keep the transmitted angle and simply rewrite Version 1 using (3.1.18) as

Version 3:

$$
R=\frac{\rho\cos\theta_i-K\cos\theta_t}{\rho\cos\theta_i+K\ \cos\theta_t} , \tag{3.1.21}
$$

and

$$
T=\frac{2\cos\theta_i}{\rho\cos\theta_i+K\cos\theta_t} . \tag{3.1.22}
$$

We note that we always have the constraint which follows from (3.1.13) that

$$
1+R=\rho T . \tag{3.1.23}
$$

If we replace K in Version 3 using Snell's law (3.1.9) we get

Version 4:

$$
R=\frac{\rho\sin\theta_t\cos\theta_i-\cos\theta_t\sin\theta_i}{\rho\sin\theta_t\cos\theta_i+\cos\theta_t\sin\theta_i} , \tag{3.1.24}
$$

and

$$
T=\frac{2\cos\theta_i\sin\theta_t}{\rho\sin\theta_t\cos\theta_i+\cos\theta_t\sin\theta_i} . \tag{3.1.25}
$$

If we define the bulk moduli $b_j=\rho_j c_j^2$ of the two media and the ratio of the bulk moduli as $b=b_2/b_1$ then (3.1.15) can be rewritten as

$$R = \frac{\left\{ \frac{\rho_2 k_1^2}{\rho_1 k_2^2} \frac{k_2}{k_1} cos\theta_i - cos\theta_t \right\}}{\left\{ \frac{\rho_2 k_1^2}{\rho_1 k_2^2} \frac{k_2}{k_1} \cos\theta_i + \cos\theta_t \right\}}$$

and substituting Snell's law for k_2/k_1 we get

Version 5:

$$R = \frac{b \sin\theta_i \cos\theta_i - \sin\theta_t \cos\theta_t}{b \sin\theta_i \cos\theta_i + \sin\theta_t \cos\theta_t} \quad , \qquad (3.1.26)$$

and

$$T = \frac{2\text{ctn}\theta_i \sin^2\theta_t}{b \sin\theta_i \cos\theta_i + \sin\theta_t \cos\theta_t} \quad , \qquad (3.1.27)$$

or using the result

$$2 \sin\theta \cos\theta = \sin 2\theta \quad ,$$

Version 6:

$$R = \frac{b \sin 2\theta_i - \sin 2\theta_t}{b \sin 2\theta_i + \sin 2\theta_t} \quad , \qquad (3.1.28)$$

and

$$T = \frac{2 \text{ ctn } \theta_i \sin^2\theta_t}{b \sin 2\theta_i + \sin 2\theta_t} \quad . \qquad (3.1.29)$$

3.2 Examples

In this section we list some examples of special cases of our reflection and transmission coefficients and point out the analogous electromagnetic coefficients.

Example 1: $\rho_2 = \rho_1$, Equal Densities

From Version 4 of Sect. 3.1 we can write

$$R = \frac{\sin(\theta_t - \theta_i)}{\sin(\theta_t + \theta_i)} \quad , \qquad (3.2.1)$$

and

$$T = \frac{2 \cos\theta_i \sin\theta_t}{\sin(\theta_t + \theta_i)} \quad . \qquad (3.2.2)$$

These are just the Fresnel reflection and transmission coefficients (of light) polarized *perpendicular* ($\boldsymbol{E} = \hat{y}E$ for the electric field) to the plane of incidence of an interface separating media of different dielectric constant but the same magnetic permeability.

Example 2: $b_2 = b_1$, Equal Bulk Moduli

From Version 6 of Sect. 3.1 and the tangent identities

$$\frac{\sin x \pm \sin y}{\cos x + \cos y} = \tan\left[\frac{1}{2}(x \pm y)\right]$$

we can write

$$R = \frac{\tan(\theta_i - \theta_t)}{\tan(\theta_i + \theta_t)} \ , \tag{3.2.3}$$

and

$$T = \frac{c_2}{c_1} \frac{2\cos\theta_i \sin\theta_t}{\sin(\theta_i + \theta_t)\cos(\theta_i - \theta_t)} \ , \tag{3.2.4}$$

which are the Fresnel reflection and transmission coefficients (of light) polarized *parallel* ($\boldsymbol{E}$ in the $x - z$ plane) to the plane of incidence for an interface again separating media of different dielectric constants and the same magnetic permeability.

Example 3: $\rho_2 c_2 = \rho_1 c_1$, Equal Normal Impedances

From Version 3 of Sect. 3.1 we can write

$$R = \frac{\cos\theta_i - \cos\theta_t}{\cos\theta_i + \cos\theta_t} \ , \tag{3.2.5}$$

and

$$T = \frac{c_2}{c_1} \frac{2\cos\theta_i}{\cos\theta_i + \cos\theta_t} \ . \tag{3.2.6}$$

3.3 Interface Values and Impedances

At the interface we have the field and normal derivative values $\psi_1(x,0)$ and $\partial\psi_1(x,0)/\partial z$. From (3.1.11) we can write (evaluating ψ_2)

$$\begin{aligned} \psi_1(x,0) &= \rho A_t e^{ixk_2 \sin\theta_t} \\ &= \rho A_i T e^{ixk_1 sin\theta_i} \\ &= \frac{2A_i\rho\cos\theta_i}{\rho\cos\theta_i + K\cos\theta_t} e^{ixk_1 \sin\theta_i} \ , \end{aligned} \tag{3.3.1}$$

where we have used Snell's law and (3.1.22) for T. Similarly from (3.1.12) we get

$$\begin{aligned}\frac{\partial\psi_1}{\partial z}(x,0) &= -ik_2\cos\theta_t A_t e^{ixk_2\sin\theta_t} \\ &= -i\,A_i k_2\cos\theta_t T e^{ixk_1\sin\theta_i} \\ &= \frac{-2i\,A_i k_2\cos\theta_i\cos\theta_t}{\rho\cos\theta_i + K\cos\theta_t}e^{ixk_1\sin\theta_i} \quad . \qquad (3.3.2)\end{aligned}$$

We have two special cases. The *Dirichlet (Soft) Boundary Condition* (with $\rho_2 = 0$) specifies that

$$\psi_1(x,0) = 0\ , \quad R = -1\ , \qquad (3.3.3)$$

so that

$$\frac{\partial\psi_1}{\partial z}(x,0) = -2iA_i k_1\cos\theta_i e^{ik_1x\sin\theta_i}\ ,$$

and

$$T = 2\cos\theta_i/K\cos\theta_t\ .$$

(Note that in our notation $1 + R = \rho T$, so that for the soft boundary, ρT vanishes not just T). The second case is the *Neumann (Hard) Boundary Condition* (with $\rho_2 = \infty$) where

$$\frac{\partial\psi_1}{\partial z}(x,0) = 0\ , \quad R = 1\ , \quad T = 0\ , \qquad (3.3.4)$$

so that

$$\psi_1(x,0) = 2A_i e^{ik_1x\sin\theta_i}\ .$$

Note that for both boundary conditions, the remaining non-zero surface field contains a factor 2. For the Dirichlet problem $\psi_1 = 0$ and $\partial\psi_1/\partial z$ on the boundary is twice the corresponding value of the z-derivative of the incident field on the boundary. For the Neumann problem, $\partial\psi_1/\partial z$ vanishes and ψ_1 doubles.

If we define the *medium (boundary) impedance* (for plane waves, derived below) as

$$Z_j = \rho_j c_j/\cos\theta_j \quad (\theta_1 = \theta_i\ ,\ \ \theta_2 = \theta_t)\ , \qquad (3.3.5)$$

then the reflection coefficient can be written from (3.1.21) as

$$R = \frac{Z_2 - Z_1}{Z_2 + Z_1}\ , \qquad (3.3.6)$$

which is similar in form to the reflection coefficient for a one-dimensional transmission line. Alternatively we can define the *normal (wall) impedance*

$$z_j = \rho_j c_j \ , \tag{3.3.7}$$

and the surface impedance

$$\beta = z_1/z_2 \ , \tag{3.3.8}$$

to write the reflection coefficient as (from (3.1.21))

$$R = \frac{\cos\theta_i - \beta\cos\theta_i}{\cos\theta_i + \beta\cos\theta_t} \ . \tag{3.3.9}$$

The boundary impedance defined above can be derived from our definition of impedance in Chap. 1 as the ratio of pressure to normal velocity

$$Z = p/v_n \ . \tag{3.3.10}$$

Using the specific forms for the surface fields this can be written as (the normal is in the negative z-direction)

$$Z = -i\omega\rho_1\psi_1(x,0)\left[\frac{\partial\psi_1}{\partial z}(x,0)\right]^{-1} \ . \tag{3.3.11}$$

From (3.3.1) and (3.3.2) we thus get

$$Z = \rho_2 c_2/\cos\theta_t \ , \tag{3.3.12}$$

which is our definition above. Note that because we used the fields in the lower region to evaluate (3.3.1) and (3.3.2) we actually derived Z_2. The overall sign of Z depends on the signs of the harmonic time convention and the direction of the surface normal.

3.4 Impedance Boundary Condition

In the previous section, the reflection coefficient was re-expressed in other forms using several definitions of impedance. We stress that this was merely the re-expression of the *exact* reflection coefficient derived for plane wave scattering from a planar interface using *two* independent continuity conditions, those for pressure and normal velocity, at the interface. The derivation of medium impedance for example used the surface values of field and normal derivative which followed from these *two* continuity conditions.

Now suppose we say that we know only one boundary condition at the interface, the impedance. That is, we are given Z where

$$Z = p(x,0)/v_n(x,0) \ . \tag{3.4.1}$$

We re-express this using the fields in region 1 to get (3.3.11). If we assume the representation

$$\begin{aligned}\psi_1(x,z) &= A_i \exp\Big\{ik_1\left[x\sin\theta_i - z\cos\theta_i\right]\Big\} \\ &\quad + A_r \exp\Big\{ik_1\left[x\sin\theta_i + z\cos\theta_i\right]\Big\} \ ,\end{aligned} \tag{3.4.2}$$

where we have assumed specular scattering ($\theta_r = \theta_i$) but where the ratio $R = A_r/A_i$ is unknown, substitute this in (3.3.11), and solve for R we get

$$Z = \frac{\rho_1 c_1}{\cos\theta_i} \ \frac{1+R}{1-R} \ , \tag{3.4.3}$$

where, using the definition of medium impedance

$$R = \frac{Z - Z_1}{Z + Z_1} \ . \tag{3.4.4}$$

This appears to be in the form in Sec. 3.3, but the second part in the impedance boundary condition assumption is to choose for Z the normal impedance of the second medium, $\rho_2 c_2$, rather than the medium impedance. Then R becomes

$$R = \frac{\cos\theta_i - \beta}{\cos\theta_i + \beta} \ , \tag{3.4.5}$$

which is similar to (3.3.9), but with the transmission angle $\theta_t = 0$. This is only consistent with the solution in region 2 of the form

$$\psi_2(x,z) = A_t \exp(-ik_2 z) \ . \tag{3.4.6}$$

The solutions (3.4.2) and (3.4.6) cannot satisfy continuity conditions everywhere on the surface because Snell's law is violated. Equivalently, we have no concept of translational invariance, i.e. the only position where the continuity conditions are satisfied is at $x = 0$, or if in some way A_t is a function of x.[1] From the solution (3.4.6), the particle motion and propagation direction are normal to the boundary, and independent of adjacent points on the boundary. Physically this is equivalent to an absorbing medium where k_2 has a positive imaginary part (and z is negative).

[1] The latter, however, changes the differential equation satisfied by the transmitted field in that we must now accompany it with a differential equation on A_t.

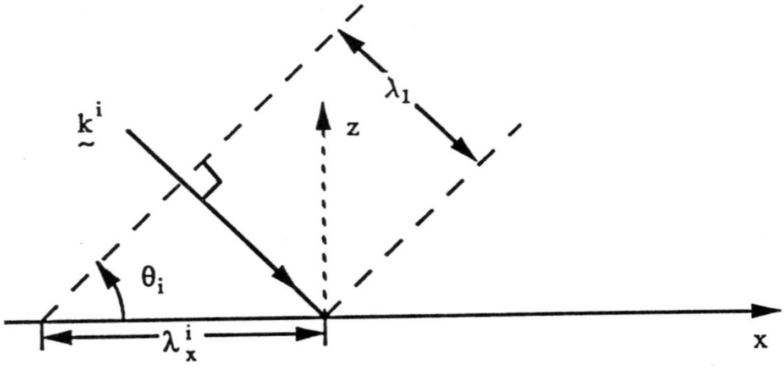

Fig. 3.2. Incident wave front projection on the planar boundary, $\lambda_x^i = \lambda_1/\sin\theta_i$.

3.5 Wave Fronts and Ray Parameter

The incident wave can be written as

$$\psi_i(\boldsymbol{x}) = A_i \exp\left(i\boldsymbol{k}^i \cdot \boldsymbol{x}\right) \quad , \tag{3.5.1}$$

where $\boldsymbol{x} = (x, 0, z)$ and

$$\boldsymbol{k}^i = (k_1 \sin\theta_i, 0, -k_1 \cos\theta_i) \quad , \tag{3.5.2}$$

with $k_1 = 2\pi/\lambda_1$. The projection of the wavefront on the planar boundary is illustrated in Fig. 3.2.

The reflected wave can be written as

$$\psi_r(\boldsymbol{x}) = A_r \exp(i\boldsymbol{k}^r \cdot \boldsymbol{x}) \quad , \tag{3.5.3}$$

where

$$\boldsymbol{k}^r = (k_1 \sin\theta_r, 0, k_1 \cos\theta_r) \quad . \tag{3.5.4}$$

The projection of its wavefront on the boundary is illustrated in Fig. 3.3.

If we require that the incident and reflected wave front projections are equal,

$$\lambda_x^i = \lambda_x^r \quad . \tag{3.5.5}$$

This implies $\theta_i = \theta_r$. The general principal involved in this is the continuity of the projection of the phase wavefront *along* the boundary.

The transmitted field can be represented as

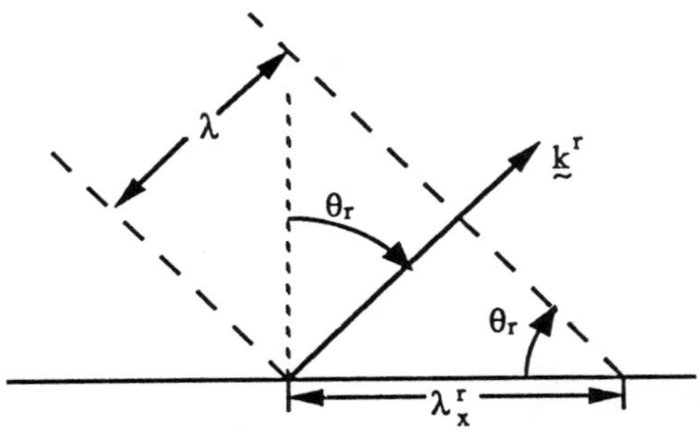

Fig. 3.3. Reflected wave front projection on the planar boundary, $\lambda_x^r = \lambda_1/\sin\theta_r$.

$$\psi_t(\boldsymbol{x}) = A_t \exp(i\boldsymbol{k}^t \cdot \boldsymbol{x}) \ , \tag{3.5.6}$$

where

$$\boldsymbol{k}^t = (k_2 \sin\theta_t, 0, -k_2\cos\theta_t) \ . \tag{3.5.7}$$

The projections of the incident and transmitted wavefronts are illustrated in Fig. 3.4.

The general principal involved here is the continuity of the projection of the phase wavefront *across* the boundary. This is

$$\lambda_x^i = \lambda_x^t \ , \tag{3.5.8}$$

which is

$$\frac{\lambda_1}{\sin\theta_i} = \frac{\lambda_2}{\sin\theta_t} \ . \tag{3.5.9}$$

This is just Snell's law for transmission. The continuity of the wave front projections is an alternate way to view the angle restrictions which apply to plane wave scattering from a flat interface.

An alternative way to understand this is to define the *ray parameter*

$$p_j = \sin\theta_j / c_j \ . \tag{3.5.10}$$

The angle restrictions then follow from the continuity of the ray parameter

$$p_1^i = p_1^r = p_2^t \ . \tag{3.5.11}$$

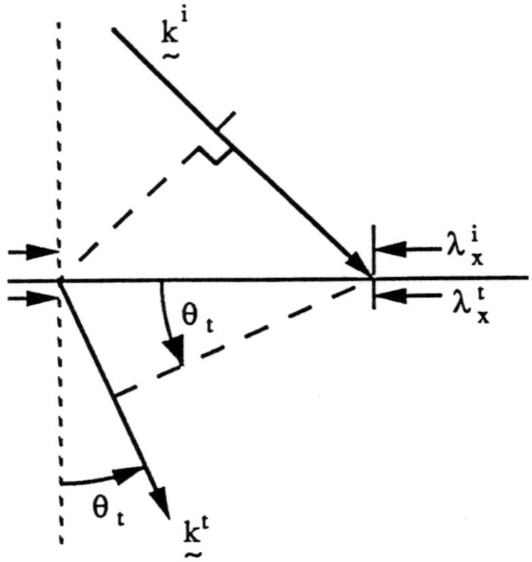

Fig. 3.4. Incident (λ_x^i) and transmitted λ_x^t wave front projections on the planar boundary.

3.6 Total Transmission

The reflection coefficient (3.1.19) vanishes for the case of total transmission. This can occur at an angle $\theta_i = \theta_B$ (the Brewster angle) when the numerator of (3.1.19) is zero. This is

$$\rho \cos\theta_B - (K^2 - \sin^2\theta_B)^{1/2} = 0 \ . \tag{3.6.1}$$

The solution of (3.6.1) yields

$$\sin\theta_B = \left(\frac{\rho^2 - K^2}{\rho^2 - 1}\right)^{1/2} \ . \tag{3.6.2}$$

The angle θ_B is thus a real angle provided

$$0 < \frac{\rho^2 - K^2}{\rho^2 - 1} < 1 \ . \tag{3.6.3}$$

There are two possible cases where this can occur.

Case 1: $\rho > K > 1$

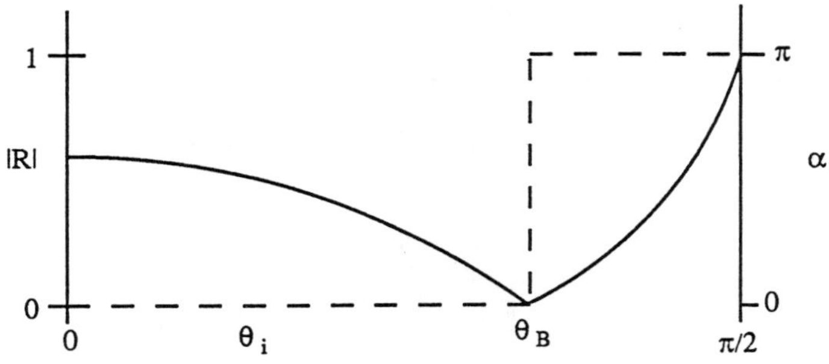

Fig. 3.5. Magnitude $|R|$ (solid line) and phase α (dashed line) of the reflection coefficient as a function of the incident angle for $\rho > K > 1$.

Since $K > 1$, it is easily seen from (3.1.19) that R is always real. If we nevertheless write it as

$$R = |R| \exp(i\alpha) \ , \tag{3.6.4}$$

it is seen that at $\theta_i = 0$, $|R| = (\rho - K)/(\rho + K)$ and $\alpha = 0$. Similarly at $\theta_i = \theta_B$, $|R| = 0$ and $\alpha = 0$, and at $\theta_i = \pi/2$, $|R| = 1$, $R = -1$, and $\alpha = \pi$. A sketch of $|R|$ and α is illustrated in Fig. 3.5.

Case 2: $\rho < K < 1$

Since $K < 1$, it can be seen from (3.1.19) that R can be complex. Define the critical angle θ_C as that angle for which the square root in (3.1.19) is zero, i.e.

$$\sin\theta_C = K = c_1/c_2 \ .$$

We have that $\theta_B < \theta_C$. This is because $\sin^2\theta_B = (K^2 - \rho^2)/(1 - \rho^2)$ and it is true that

$$\frac{K^2 - \rho^2}{1 - \rho^2} < K^2,$$

for $K < 1$. $|R|$ and α for this case are illustrated in Fig. 3.6. Simple points in Fig. 3.6 are

(a) $\theta_i = 0, \quad R = (\rho - K)/(\rho + K) < 0, \quad \alpha = \pi$

(b) $\theta_i = \theta_B, \quad R = 0, \quad \alpha = 0$

(c) $\theta_i = \theta_C, \quad R = 1, \quad \alpha = 0$

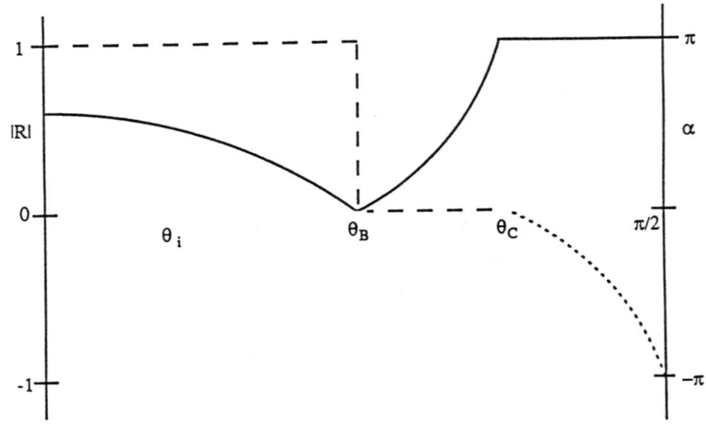

Fig. 3.6. Magnitude $|R|$ (solid line) and phase α (dashed line) of the reflection coefficient for $\rho < K < 1$.

$$\textbf{(d)}\ \theta_i > \theta_C, \quad |R| = 1, \quad \alpha = -2\tan^{-1}\left\{\frac{\left[\sin^2\theta_i - K^2\right]^{1/2}}{\rho\cos\theta_i}\right\}$$

where, as $\theta_i \to \pi/2$, $\alpha \to -\pi$.

We consider some examples. For the air-water interface we have parameters illustrated in Fig. 3.7.

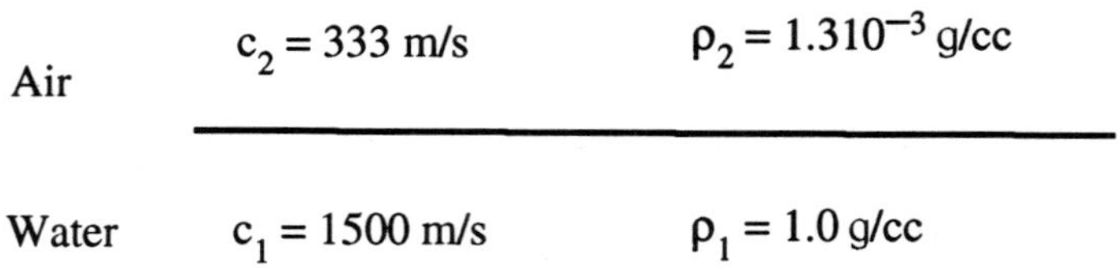

Fig. 3.7. Typical parameters for the air-water interface.

Eg. 1 For a source in the water (region 1) we have $\rho = \rho_2/\rho_1 = 1.3 \cdot 10^{-3}$ and $K = k_2/k_1 = c_1/c_2 = 1500/333 \sim 5$. Since $\rho < 1$ and $K > 1$, θ_B is not a real angle.

Eg. 2 For a source in the air (which is now region 1 in our notation), the density and wavenumber ratios are the inverse of those in Eg. 1. Hence $\rho > 1$ and $K < 1$, and θ_B is again not a real angle.

The parameters for a water-bottom interface are illustrated in Fig. 3.8.

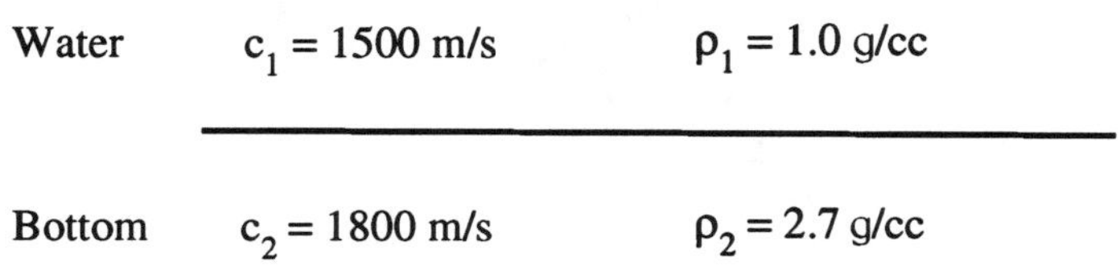

Fig. 3.8. Typical parameters for a water bottom interface where the bottom is rock.

Eg. 3 For a source in the water above a rock bottom we have $\rho = 2.7$ and $K = 1500/1800$. Thus, $\rho > 1$ and $K < 1$ and θ_B is not a real angle.

Eg. 4 For a source in the bottom the parameters are the inverse of those in Eg. 3. Hence, $\rho < 1$ and $K > 1$ so that again θ_B is not a real angle.

We see that for several common examples there is no real Brewster angle and total transmission does not occur. It is not as common a phenomenon in scalar wave theory as it is in electromagnetic theory where one of the polarizations is extinguished. A scalar example of where this could occur is the case of a muddy bottom transition layer where the sound speed in this bottom is actually lower than in the water. It is only lower for a short distance however and further reflections from a harder underlying bottom can further alter the field in the water.

3.7 Total Internal Reflection

This case occurs when we reflect off a medium which contains a faster wave speed than our source medium. The ray direction in the transmitted medium bends outward from the normal as illustrated in Fig. 3.9. Since $K < 1$, this case admits a critical angle.

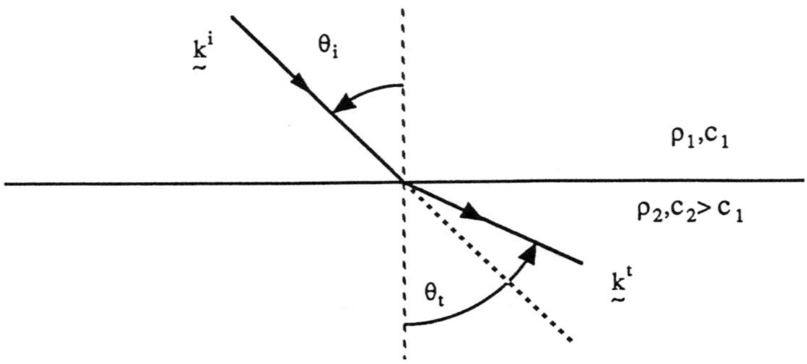

Fig. 3.9. Ray paths for scattering from a slower to a faster medium.

When $\theta_i = \theta_c$, $R = 1$ and the field in region 1 is given by (3.1.2) - (3.1.4) evaluated at θ_c which yields

$$\psi_1(x, z) = 2A_i \cos(k_1 z \cos\theta_c) \exp(ik_1 x \sin\theta_c) \quad , \tag{3.7.1}$$

which consists of a standing wave in the z-direction or equivalently a cosine envelope.

The transmission coefficient at the critical angle is seen to equal $2/\rho$ from (3.1.20). Since $\theta_t = \pi/2$ the field in region 2 is given by

$$\psi_2(x, z) = (2/\rho) \exp(ik_2 x) \quad , \tag{3.7.2}$$

which is a wave propagating in the positive x-direction along the boundary. For (3.5.7) we have that $\boldsymbol{k}^t = k_2(1, 0, 0)$. This is illustrated in Fig. 3.10. Various terms can be applied to this wave. It can be called a lateral wave, an interface wave, a trace wave, a surface wave, or a refracted arrival (seismology). The wave propagates along the boundary

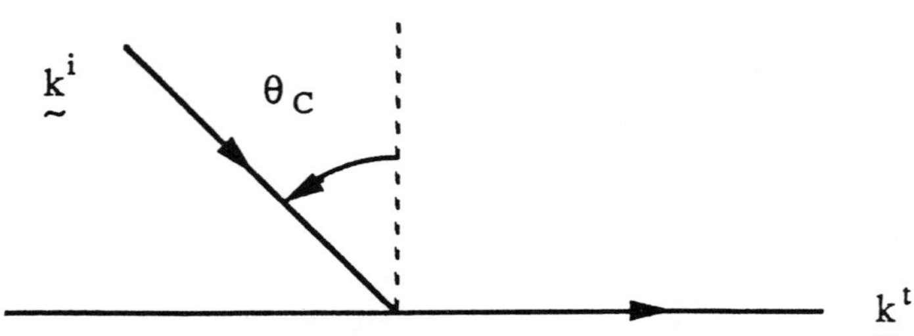

Fig. 3.10. Incident wave coupled directly into the boundary at critical angle incidence.

with speed c_2 the same as any wave in medium 2 except that it is restricted to move along the boundary. Strictly speaking it is not a true surface wave since we have no decay away from the boundary. This occurs for incident angles greater than critical which are discussed next. Lateral waves are considered further in Chap. 4.

For *incidence angles greater than critical* the reflection coefficient (3.1.19) is written as

$$R = \frac{\rho\cos\theta_i - i(\sin^2\theta_i - K^2)^{1/2}}{\rho\cos\theta_i + i(\sin^2\theta_i - K^2)^{1/2}} \; . \tag{3.7.3}$$

Using (3.6.4) we see that $|R| = 1$ and

$$\alpha = -2\tan^{-1}\left\{\frac{(\sin^2\theta_i - K^2)^{1/2}}{\rho\cos\theta_i}\right\} \; .$$

The field in region 1 is thus written as

$$\begin{aligned} \psi_1(x,z) &= A_i\Big\{\exp\Big[ik_1(x\sin\theta_i - z\cos\theta_i)\Big] \\ &\qquad + \exp\Big[ik_1(x\sin\theta_i + z\cos\theta_i) + i\alpha\Big]\Big\} \\ &= 2A_i\cos(k_1 z\cos\theta_i + \alpha/2)\exp(ik_1 x\sin\theta_i + \alpha/2) \; , \end{aligned} \tag{3.7.4}$$

which again yields a cosine envelope but now phase shifted by $\alpha/2$ due to the reflection coefficient. The field in region 2 is more complicated. Since $\theta_i > \theta_c$, we have that $\sin\theta_i > K$. It follows from Snell's law that $\sin\theta_t > 1$. If we represent θ_t in terms of real and imaginary parts

$$\theta_t = \theta_{tr} + i\theta_{ti} \; , \tag{3.7.5}$$

and use the fact that $\theta_{tr} = \pi/2$ (the transmitted wave travels along the boundary), we have

$$\sin\theta_t = \cosh\theta_{ti} \; , \quad \cos\theta_t = -i\sinh\theta_{ti} \; . \tag{3.7.6}$$

The latter implies that $\theta_{ti} < 0$. Further we have that

$$\cos\theta_t = i(\sin^2\theta_t - 1)^{1/2} \; , \tag{3.7.7}$$

and from Snell's law

$$\sin\theta_t = K^{-1}\sin\theta_i \; , \tag{3.7.8}$$

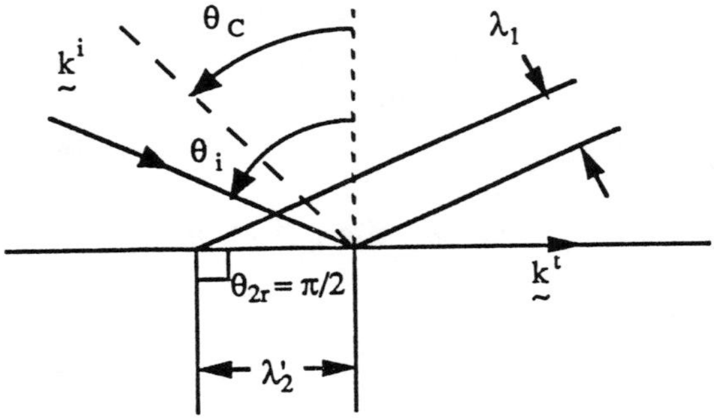

Fig. 3.11. Effective wavelength λ_2' of the transmitted field for incidence angles greater than critical.

so that

$$\cos\theta_t = iK^{-1}(\sin^2\theta_i - K^2)^{1/2} \ . \tag{3.7.9}$$

Using (3.1.23) the transmission coefficient is

$$T = \rho^{-1}(1 + e^{i\alpha}) \ , \tag{3.7.10}$$

so that the field in region 2

$$\psi_2(x,z) = T\exp\Big[ik_2(x\sin\theta_t - z\cos\theta_t)\Big] \ , \tag{3.7.11}$$

can be rewritten as

$$\psi_2(x,z) = \rho^{-1}(1 + e^{i\alpha}) \cdot \exp\Big\{ik_1x\sin\theta_i + k_1z(\sin^2\theta_i - K^2)^{1/2}\Big\} \ , \tag{3.7.12}$$

which is in the form of an inhomogeneous plane wave which decays into region 2 ($z < 0$). The wave in region 2 propagates along the boundary with an effective speed c_2' where

$$k_2' = k_1\sin\theta_i = k_2\sin\theta_t$$

and

$$c_2' = \frac{c_2}{\sin\theta_t} < c_2 \tag{3.7.13}$$

where the latter inequality follows since $\sin\theta_t > 1$. The effective wavelength λ_2' is also reduced by a factor $\sin\theta_t$,

$$\lambda_2' = \frac{\lambda_2}{\sin\theta_i} = \frac{\lambda_2}{\sin\theta_t} < \lambda_2 \ . \tag{3.7.14}$$

We illustrate the latter in Fig. 3.11. The transmitted field thus propagates along the boundary, decays away from it (into region 2) and has an effective wave speed and wavelength reduced from those of homogeneous waves in region 2.

3.8 Effect of Surface Roughness

We have been considering the scattering of a plane wave from a flat interface $z = 0$. If instead the surface is rough in one dimension, $z = h(x)$, the scattered wave is modified by a change in phase. Consider the two plane waves illustrated in Fig. 3.12. Plane wave 1 scatters specularly from the $z = 0$ surface as if there were no roughness. Plane wave 2 scatters from the same x-point but from a rough surface $h(x)$ into an assumed specular direction. It experiences an additional spatial path length 2δ, and an additional phase change

$$\begin{aligned} \Delta\beta &= k2\delta \\ &= 2kh\cos\theta_i \ . \end{aligned} \tag{3.8.1}$$

That is, the reflected field now becomes

$$\psi_r(x,z) \rightarrow \psi_r(x,z)e^{i\Delta\beta} \ , \tag{3.8.2}$$

with the (assumed) same reflection coefficient. The criterion that the surface is "smooth" essentially states that these two plane waves should not get so far out of phase that they destructively interfere with each other. We assume the surface is smooth if

$$\Delta\beta < \pi/2 \ , \tag{3.8.3}$$

or, equivalently if

$$\frac{h}{\lambda}\cos\theta_i < \frac{1}{8} \ . \tag{3.8.4}$$

Otherwise the surface is considered "rough." This criterion is due to Rayleigh. For an incident plane wave

$$e^{i\boldsymbol{k}^i\cdot\boldsymbol{x}} \ , \tag{3.8.5}$$

where $\boldsymbol{k}^i = (k_x^i, 0, k_z^i)$ with $k_x^i = k\sin\theta_i$ and $k_z^i = -k\cos\theta_i$, and a scattered plane wave (in the specular direction)

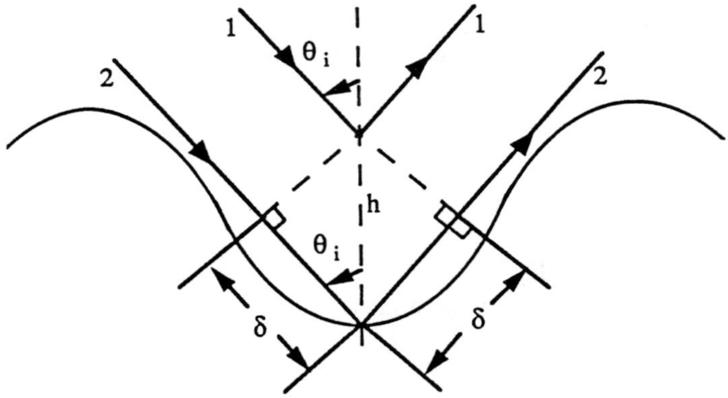

Fig. 3.12. Plane wave scattering from a surface roughness element of height h. Plane wave 1 scatters as if there were no rough surface. Plane wave 2 scatters from the roughness but still in the specular direction. The path difference is 2δ.

$$e^{i\boldsymbol{k}^s\cdot\boldsymbol{x}} \quad , \tag{3.8.6}$$

where $\boldsymbol{k}^s = (k_x^s, 0, k_z^s)$ with $k_x^s = k\sin\theta_i$ and $k_z^s = k\cos\theta_i$, we can rewrite the phase shift as

$$\Delta\beta = (k_z^s - k_z^i)h \quad . \tag{3.8.7}$$

More generally, if the plane wave scatters from two actual surface points separated by a distance Δx, there is an additional phase term of the form

$$(k_x^s - k_x^i)\Delta x \quad . \tag{3.8.8}$$

Here the $k_x^i\Delta x$ term is the phase path which would be taken by the incident plane wave if it were translated by Δx and scattered specularly. The term $k_x^s\Delta x$ is the phase component of the true scattered wave that scatters specularly off the surface (locally) but not necessarily in the overall specular direction since the real surface point is tilted from the horizontal. If the two surface points are also separated in height by Δh, then the total phase change is

$$\Delta\beta = (k_z^s - k_z^i)\Delta h + (k_x^s - k_x^i)\Delta x \quad , \tag{3.8.9}$$

or by

$$\Delta\beta = (\boldsymbol{k}^s - \boldsymbol{k}^i)\cdot(\Delta x, \Delta h) \quad , \tag{3.8.10}$$

that is, by the scalar product of the change in the wavevector $\Delta \boldsymbol{k} = \boldsymbol{k}^s - \boldsymbol{k}^i$ with the coordinate difference vector $(\Delta x, \Delta h)$. This argument can be extended to two-dimensional surfaces $h(x, y)$ as

$$\Delta\beta = (\boldsymbol{k}^s - \boldsymbol{k}^i) \cdot (\Delta x, \Delta y, \Delta h), \tag{3.8.11}$$

with appropriately defined incident and scattered wavevectors in three dimensions.

As a simple example of the effect of roughness, consider the case when the surface height can be treated as a random variable. The reflected field is (in the specular direction)

$$\psi_r(x, z) = A_r \exp\left\{ik(x \sin\theta_i + z\cos\theta_i) + i\Delta\beta\right\} \ . \tag{3.8.12}$$

Its ensemble average (or mean), denoted by the symbol E, is found using the relation

$$E \exp(i\Delta\beta) = \int_{-\infty}^{\infty} \exp(i\Delta\beta) P(h) dh \ , \tag{3.8.13}$$

where $P(h)dh$ is the probability of recurrence of a wave reflected from the surface at angle θ_i and at that point on the surface with a height between h and $h + dh$. For simplicity we choose $\Delta\beta = 2kh\cos\theta_i$, and P to be a Gaussian

$$P(h) = \frac{1}{\sigma\sqrt{2\pi}} \exp(-h^2/2\sigma^2) \ , \tag{3.8.14}$$

where σ is the *rms* surface height, then we get

$$E \exp(i\Delta\beta) = \exp(-2\Sigma^2) \ , \tag{3.8.15}$$

where

$$\Sigma = k\sigma\cos\theta_i \ , \tag{3.8.16}$$

is the Rayleigh roughness parameter. The resulting mean field in the specular direction $E\psi_r(x, z)$ has a decay term. Equivalently we can say that the *mean reflected field remains in the specular direction but its reflection coefficient R is replaced by*

$$R \exp\left(-2\Sigma^2\right) \ . \tag{3.8.17}$$

This is the simplest possible modification of the specularly reflected wave due to surface roughness. The limits of its validity and a discussion of more complicated models can be found in Ref. [3.4].

4. Spherical Waves Scattering from Planar Boundaries

In this chapter we consider the scattering of a spherical wave from a planar boundary separating two different fluids with densities ρ_j and sound speeds $c_j (j = 1, 2)$. This is the same geometry considered in Chap. 3 with the additional complication that all incidence angles are now possible. For the point source located at $\boldsymbol{x}'(z' > 0)$ in the upper fluid (fluid 1) and the receiver at $\boldsymbol{x}$ which could occur in either fluid we have the geometry illustrated in Fig. 4.1.

4.1 Weyl and Sommerfeld Representations

We previously discussed the Weyl and Sommerfeld representations in Chap. 2. These were representations of the free-space Green's function or point source solution of the Helmholtz equation. The z-direction is the characteristic direction and the Weyl representation of $G^{(3)}$ is

$$G^{(3)}(\boldsymbol{x}, \boldsymbol{x}', k_0) = \frac{\pi i}{[2\pi]^3} \int\int e^{i\boldsymbol{k}_t \cdot (\boldsymbol{x}_t - \boldsymbol{x}_t')} \frac{e^{iK|z-z'|}}{K} d\boldsymbol{k}_t \ , \tag{4.1.1}$$

where $K = (k_0^2 - k_t^2)^{1/2}$, $k_t = (k_x^2 + k_y^2)^{1/2}$ and $\operatorname{Im} k_0 > 0$. We assume cylindrical symmetry so our results will depend on only the radius in the x-y plane, $r = |\boldsymbol{x}_t - \boldsymbol{x}_t'|$. We integrate (4.1.1) over angle to derive the Sommerfeld representation

$$G^{(3)}(\boldsymbol{x}, \boldsymbol{x}', k_0) = \frac{i}{4\pi} \int_0^\infty \frac{J_0(k_t r) e^{iK|z-z'|}}{K} k_t d\boldsymbol{k}_t \ . \tag{4.1.2}$$

This is the representation of a point source which will prove to be useful for the case of a cylindrically symmetric physical geometry. Explicitly we still have

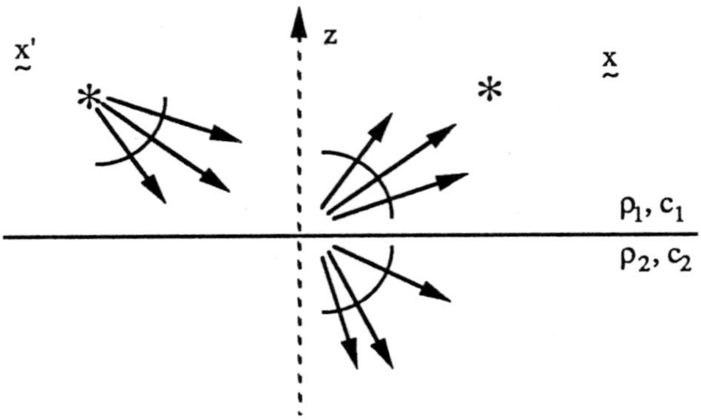

Fig. 4.1. Spherical wave scattering and transmission at a flat interface between two semi-infinite fluids.

$$G^{(3)}(\boldsymbol{x}, \boldsymbol{x}', k_0) = \frac{e^{ik_0 R}}{4\pi R}; \qquad R = \left[r^2 + (z - z')^2\right]^{1/2} . \tag{4.1.3}$$

The representation (4.1.2) involves integration over all angles. We can still parametrize k_x and k_y as

$$k_x = \frac{\omega}{c} \sin\theta \cos\phi \, ; \qquad k_y = \frac{\omega}{c} \sin\theta \sin\phi \quad ,$$

so that

$$k_t = \frac{\omega}{c} \sin\theta \doteq \omega p \tag{4.1.4}$$

where we define the *ray parameter* p. Using $k_0 = \omega/c$ we can also define

$$K = (k_0^2 - k_t^2)^{1/2} = \omega(c^{-2} - p^2)^{1/2} \doteq \omega\xi \ . \tag{4.1.5}$$

The definitions (4.1.4) and (4.1.5) separate out the frequency parameter ω. Using these we write (4.1.2) as

$$G^{(3)}(\boldsymbol{x}, \boldsymbol{x}', k_0) = \frac{i\omega}{4\pi} \int_0^\infty \frac{J_0(\omega p r) e^{i\omega\xi|z-z'|}}{\xi} p dp \ . \tag{4.1.6}$$

The ray parameter p is also referred to as the *horizontal slowness*, and the parameter $\xi = (c^{-2} - p^2)^{1/2}$ is called the *vertical slowness.* They are related as

$$p^2 + \xi^2 = c^{-2} \ . \tag{4.1.7}$$

Note that the representation (4.1.6) involves an integration over all possible ray parameters p.

4.2 Field Representations

The point source incident in region 1 has an arbitrary amplitude A. This incident field ψ_i satisfies the differential equation

$$\left(\nabla^2 + k_1^2\right) \psi_i(r, z, z') = -A\frac{\delta(r)\delta(z - z')}{2\pi r} \ . \qquad (4.2.1)$$

The solution is the angle-independent point source

$$\psi_i(r, z, z') = AG^{(3)}(\boldsymbol{x}, \boldsymbol{x}', k_1) \ , \qquad (4.2.2)$$

which is given by the Sommerfeld representation with $\xi_1 = (c_1^{-2} - p^2)^{1/2}$ as

$$\psi_i(r, z, z') = \frac{i\omega}{4\pi} A \int_0^\infty \frac{J_0(\omega pr)e^{i\omega\xi_1|z-z'|}}{\xi_1} pdp \ . \qquad (4.2.3)$$

The reflected field ψ_r satisfies the (sourceless) differential equation (for $z > 0$)

$$\left(\nabla^2 + k_1^2\right) \psi_r(r, z, z') = 0 \ , \qquad (4.2.4)$$

and in analogy with (4.2.3) we write it as

$$\psi_r(r, z, z') = \frac{i\omega}{4\pi} \int_0^\infty \frac{B(p)J_0(\omega pr)e^{i\omega\xi_1(z+z')}}{\xi_1} pdp \ , \qquad (4.2.5)$$

where $B(p)$ is an arbitrary amplitude modulating function of the ray parameter which must be found using the boundary conditions below. It is essentially the reflection coefficient which for spherical waves now depends explicitly on the variable ray parameter. The representation (4.2.5) can also be thought of as the solution for $z > 0$ of an image point source. The total field in region 1 is

$$\psi_1(r, z, z') = \psi_i(r, z, z') + \psi_r(r, z, z') \ . \qquad (4.2.6)$$

The depth dependent phase term in (4.2.5) can be interpreted as the incident (image) ray travelling a distance z' and the scattered ray a distance z. Both propagations occur in region 1.

In region 2 the transmitted velocity potential ψ_t satisfies the differential equation

$$(\nabla^2 + k_2^2)\psi_t(r, z, z') = 0 \ . \qquad (4.2.7)$$

The solution of (4.2.7), again in analogy with (4.2.3), is written as

$$\psi_t(r,z,z') = \frac{i\omega}{4\pi}\int_0^\infty \frac{C(p)J_0(\omega pr)e^{i\omega(-\xi_2 z+\xi_1 z')}}{\xi_1} pdp \ , \tag{4.2.8}$$

where $\xi_2 = (c_2^{-2} - p^2)^{1/2}$ and $C(p)$ is the unknown transmission coefficient. We have chosen to write ψ_t using the phase factor $\exp(i\omega\xi_1 z')$ and the amplitude term ξ_1^{-1} (rather than ξ_2^{-1}). This is because the transmission coefficient $C(p)$ is unknown and these factors occur in ψ_i and ψ_r when we set z on the boundary $z = 0$. They are simply additional scale factors in the transmitted field which scale out using the boundary conditions below. Here the depth dependent phase term in the integral is interpreted as an incident ray travelling a distance z' in region 1 and the transmitted ray a distance z in region 2 (propagating downwards).

The boundary conditions are the same as those in Chap. 3, continuity of pressure and normal velocity at the $z = 0$ interface. These are respectively

$$\rho_1\Big(\psi_i(r,0,z') + \psi_r(r,0,z')\Big) = \rho_2\psi_t(r,0,z') \ , \tag{4.2.9}$$

and

$$\frac{\partial}{\partial z}\Big(\psi_i(r,0,z') + \psi_r(r,0,z')\Big) = \frac{\partial}{\partial z}\psi_t(r,0,z') \ . \tag{4.2.10}$$

Substituting (4.2.3), (4.2.5), and (4.2.8) into (4.2.9) and (4.2.10) we find that the equations can be satisfied provided

$$\rho_1\Big[A + B(p)\Big] = \rho_2\, C(p) \ , \tag{4.2.11}$$

and

$$\xi_1\Big[(A - B(p)\Big] = \xi_2 C(p) \ , \tag{4.2.12}$$

whose solutions are

$$\frac{B(p)}{A} = \frac{\rho_2\xi_1 - \rho_1\xi_2}{\rho_2\xi_1 + \rho_1\xi_2} \doteq R(p) \ , \tag{4.2.13}$$

and

$$\frac{C(p)}{A} = \frac{2\rho_2\xi_1}{\rho_2\xi_1 + \rho_1\xi_2} \doteq T(p) \ , \tag{4.2.14}$$

for the reflection (R) and transmission (T) coefficients respectively. Explicitly writing out the square roots we have

$$R(p) = \frac{\rho_2(c_1^{-2} - p^2)^{1/2} - \rho_1(c_2^{-2} - p^2)^{1/2}}{\rho_2(c_1^{-2} - p^2)^{1/2} + \rho_1(c_2^{-2} - p^2)^{1/2}} \ , \tag{4.2.15}$$

and

$$T(p) = \frac{2\rho_2(c_1^{-2} - p^2)^{1/2}}{\rho_2(c_1^{-2} - p^2)^{1/2} + \rho_1(c_2^{-2} - p^2)^{1/2}} \ . \tag{4.2.16}$$

These are similar to the plane wave reflection and transmission coefficients in Chap. 1, but now for variable p.

Since both R and T are even functions of p, we can use the identity

$$J_0(\omega pr) = \frac{1}{2}\left[H_0^{(1)}(\omega pr) + H_0^{(2)}(\omega pr)\right] \ , \tag{4.2.17}$$

in (4.2.5) and (4.2.8), change variables from $p \to -p$ in the resulting $H_0^{(2)}$ integral and use the symmetry relation

$$H_0^{(2)}(-\omega pr) = -H_0^{(1)}(\omega pr) \ , \tag{4.2.18}$$

to write

$$\psi_r(r, z, z') = \frac{i\omega A}{8\pi}\int_{-\infty}^{\infty} \frac{R(p)H_o^{(1)}(\omega pr)e^{i\omega\xi_1(z+z')}}{\xi_1} pdp \ , \tag{4.2.19}$$

and

$$\psi_t(r, z, z') = \frac{i\omega A}{8\pi}\int_{-\infty}^{\infty} \frac{T(p)H_0^{(1)}(\omega pr)e^{i\omega(-\xi_2 z+\xi_1 z')}}{\xi_1} pdp \ . \tag{4.2.20}$$

We study the behavior of these two fields in the limit when the argument of the Hankel function is large. For this case we have the asymptotic expansion ($x >> 1$)

$$H_0^{(1)}(x) \sim \left(\frac{2}{\pi x}\right)^{1/2} e^{i[x-\pi/4]}\left\{1 - \frac{i}{8x} + 0\left(\frac{1}{x^2}\right)\right\} \ . \tag{4.2.21}$$

We keep only the leading order term, and (4.2.19) and (4.2.20) become

$$\psi_r(r, z, z') \sim \frac{A}{8\pi}\left(\frac{2i\omega}{\pi r}\right)^{1/2}\int_{-\infty}^{\infty} R(p)p^{1/2}\xi_1^{-1}e^{i\omega[pr+\xi_1(z+z')]}\, dp \ , \tag{4.2.22}$$

and

$$\psi_t(r, z, z') \sim \frac{A}{8\pi}\left(\frac{2i\omega}{\pi r}\right)^{1/2}\int_{-\infty}^{\infty} T(p)p^{1/2}\xi_1^{-1}\, e^{i\omega[pr-\xi_2 z+\xi_1 z']}\, dp \ . \tag{4.2.23}$$

To study the behavior of these latter integrals we first study the behavior of their integrands. We consider p to be a complex variable. Clearly using (4.2.15) and (4.2.16) these integrands have singularities of the form

(a) branch points at $p = \pm c_1^{-1}$

(b) branch points at $p = \pm c_2^{-1}$

(c) a possible pole that occurs when the denominator of $R(p)$ or $T(p)$ vanishes, i.e. at that value of p where

$$\rho_2(c_1^{-2} - p^2)^{1/2} + \rho_1(c_2^{-2} - p^2)^{1/2} = 0 \ .$$

We first study the properties of generic square root branch points of the form ξ_1 or ξ_2.

4.3 Square Root Branch Points

Here we study the analytic properties of the generic square root branch function of the form

$$\xi = (c^{-2} - p^2)^{1/2} \ . \tag{4.3.1}$$

It is clear that this function has branch points at $p = \pm c^{-1}$. The major question is how to draw the branch cuts. We determine the equation for the branch cut from the equation

$$\operatorname{Im} \xi = 0 \ . \tag{4.3.2}$$

The reason is that in the reflected field, we have a (generic) wave shape of the form

$$\exp(i\omega\xi z) \ . \tag{4.3.3}$$

Since ω and z are positive, we cannot have $\operatorname{Im} \xi < 0$ on what we call the *first* or *top* or *physical sheet* of this complex function, since this would lead to growing exponential solutions which are unphysical. *On this physical sheet we require*

$$\operatorname{Im} \xi > 0 \ , \tag{4.3.4}$$

and we now discuss how to construct the branch cuts such that this is true. The curve given by $\operatorname{Im} \xi = 0$ divides these two regions.

We write p and ξ as complex variables and consider the p-plane to be a complex plane. We have

$$p = p_r + i\, p_i \ , \tag{4.3.5}$$

and

$$\xi = (c^{-2} - p^2)^{1/2} ,$$

$$\xi = \xi_r + i\,\xi_i \ . \tag{4.3.6}$$

For simplicity we also initially consider the inverse of the speed c to have both real and imaginary parts

$$c^{-1} = a + i\delta \ \ (a > 0, \delta > 0) \ . \tag{4.3.7}$$

(This is in fact true for absorbing media.) Then

$$\xi^2 = (a + i\delta)^2 - (p_r + ip_i)^2$$

$$= a^2 - \delta^2 - p_r^2 + p_i^2 + 2i(a\delta - p_r p_i) \tag{4.3.8}$$

$$= \xi_r^2 - \xi_i^2 + 2i\,\xi_r\xi_i \ . \tag{4.3.9}$$

Equating the real and imaginary parts of (4.3.8) and (4.3.9) we have

$$\xi_r\xi_i = a\delta - p_r p_i \ , \tag{4.3.10}$$

and

$$\xi_r^2 - \xi_i^2 = a^2 - \delta^2 - p_r^2 + p_i^2 \ . \tag{4.3.11}$$

The curves on which the branches lie are

$$\operatorname{Im}\xi = \xi_i = 0 \ , \tag{4.3.12}$$

or from (4.3.10)

$$p_r p_i = a\,\delta > 0 \ , \tag{4.3.13}$$

which in the complex p-plane are the equations of two hyperbolas. In this parametrization the branch points occur at

$$p = \pm(a + i\delta) \tag{4.3.14}$$

illustrated in Fig. 4.2.

We thus have that:

(a) $\xi_i = 0$ on the hyperbolas.

(b) $\xi_i > 0$ in the rest of the plane illustrated in Fig. 4.2. This is required so that the reflected wave in (4.2.22) decays for $z > 0$ (i.e. $\xi_{1i} > 0$). The specific values of ξ_r and ξ_i are computed below.

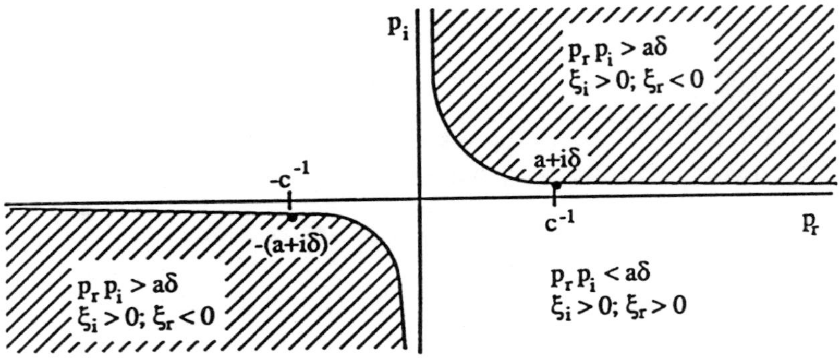

Fig. 4.2. Hyperbolae which define the locus of the branch cuts for the function ξ given by (4.3.1). The values ξ_r and ξ_i are computed in the text.

(c) The branch cuts start at $p = \pm c^{-1}$ (for $\delta = 0$). They are asymptotic to $\xi_r > 0$ in order, for example, for the phase term $\exp(i\omega\xi_1 z)$ to represent an outgoing wave for $z > 0$ in (4.2.22). On the branch lines, since $\xi_i = 0$, we have

$$\xi_r^2 = a^2 - \delta^2 - p_r^2 + p_i^2$$

from (4.3.11). If the branch lines start at $a + i\delta$ and end asymptotic to the real axis so that $p_i \to 0$ and $p_r \to \infty$, we have that at these points

$$\xi_r^2 \sim -p_r^2$$

so that ξ_r is not real. This is not true. We conclude that the *branch lines must be hyperbolas asymptotic to* $p_r = 0$.

(d) Finally, in the limit $\delta \to 0$ the hyperbolas shrink to the axes and the result is illustrated in Fig. 4.3.

Define the angles α and β in Fig. 4.3 as follows:

$$\begin{aligned} c^{-1} - p &= |c^{-1} - p| e^{i\alpha} \ , \\ &= R_+ e^{i\alpha} \ , \end{aligned} \tag{4.3.15}$$

where

$$R_+ = \left[(c^{-1} - p_r)^2 + p_i^2\right]^{1/2} \ , \tag{4.3.16}$$

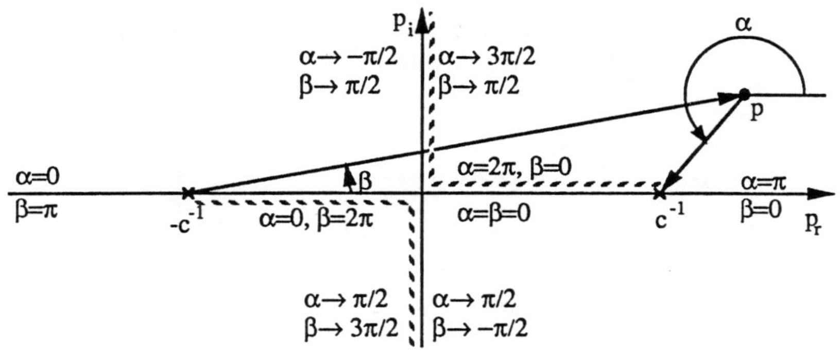

Fig. 4.3. Branch cuts starting at $\pm c^{-1}$ (slightly offset from the axes for clarity) and the corresponding angles α and β with specific examples.

and

$$\tan\alpha = p_i(p_r - c^{-1})^{-1} \quad . \tag{4.3.17}$$

We also have that

$$\begin{aligned} c^{-1} + p &= |c^{-1} + p| e^{i\beta} \quad , \\ &= R_- e^{i\beta} \quad , \end{aligned} \tag{4.3.18}$$

where

$$R_- = \left[(c^{-1} + p_r)^2 + p_i^2\right]^{1/2} \quad , \tag{4.3.19}$$

and

$$\tan\beta = p_i(c^{-1} + p_r)^{-1} \quad . \tag{4.3.20}$$

Then using (4.3.6) we get

$$\xi = (R_+ R_-)^{1/2} e^{i(\alpha+\beta)/2} \quad , \tag{4.3.21}$$

$$\xi_r = (R_+ R_-)^{1/2} \cos\left[(\alpha + \beta)/2\right] \quad , \tag{4.3.22}$$

and

$$\xi_i = (R_+ R_-)^{1/2} \sin\left[(\alpha + \beta)/2\right] \quad . \tag{4.3.23}$$

Using the angle definitions in Fig. 4.3, it can be seen that we have defined the top sheet of a Riemann surface for which $\xi_i > 0$ everywhere on the sheet. Some examples are treated in Table 4.1.

Table 4.1. Representative quadrant points and their angle definitions on the first sheet. Using (4.3.23) this illustrates that $\xi_i > 0$ everywhere on this sheet.

	POINTS	α	β	$(\alpha+\beta)/2$	$\cos\left[(\alpha+\beta)/2\right]$	$\sin\left[(\alpha+\beta)/2\right]$
1.	$p_r = 1/c$ $p_i = 2/c$	$3\pi/2$	$\pi/4$	$7\pi/8$	< 0	> 0
2.	$p_r = -1/c$ $p_i = 2/c$	$-\pi/4$	$\pi/2$	$\pi/8$	> 0	> 0
3.	$p_r = -1/c$ $p_i = -2/c$	$\pi/4$	$3\pi/2$	$7\pi/8$	< 0	> 0
4.	$p_r = 1/c$ $p_i = -2/c$	$\pi/2$	$-\pi/4$	$\pi/8$	> 0	> 0

This satisfies the radiation condition for the z-component of the reflected wave, i.e. for $\exp(i\omega\xi_1 z)$ for $z > 0$. In addition, in order to ensure physical solutions for the radial wave term $\exp(i\omega pr)$ we must require that $p_i > 0$ and this confines our integration contour to the upper half of the p-plane. Continuation of the point p through the branch cut starting at $1/c$ defines the second sheet of this Riemann surface. This type of continuation will be of interest in later sections of this chapter as well as in Chap. 5. The real and imaginary wavenumber terms can be found by replacing α by $\alpha + 2\pi$. In Table 4.2 we illustrate the angle values for representative quadrant points on the second sheet. For each of these points it is clear using (4.3.23) that $\xi_i < 0$ on this sheet. This gives rise to growing exponential solutions which are not physical.

4.4 Reflected Field

In this section we consider the asymptotic evaluation of the reflected field ψ_r given by (4.2.22). We have already used the asymptotic form of the Hankel function. We are restricting our integral to the upper half plane $p_i > 0$ for convergence in the radial phase term and to $\xi_{1i} > 0$ (physical sheet) for convergence in the depth-dependent phase terms. We previously mentioned the occurrence of other singularities in this integrand but for the moment we ignore them and evaluate ψ_r using the steepest descent method explained in Appendix 4A. The integral in the reflected field in (4.2.22) is thus of the form

$$I_1(\omega) = \int_C F(p) e^{\omega f(p)} dp \ , \tag{4.4.1}$$

where

$$f(p) = i\left[pr + \xi_1(z+z')\right] \ , \tag{4.4.2}$$

and

$$F(p) = R(p) p^{1/2}\, \xi_1^{-1} \ . \tag{4.4.3}$$

Following the presentation in Appendix 4A we can evaluate (4.4.1) for ω large using the method of steepest descent. The saddle point p_s occurs when

$$f'(p_s) = 0 \ , \tag{4.4.4}$$

or when

$$r\xi_{1s} = p_s(z+z') \ . \tag{4.4.5}$$

In terms of an angle parametrization

$$p_s = c_1^{-1} \sin\theta_s \ , \quad \xi_{1s} = c_1^{-1} cos\theta_s \ , \tag{4.4.6}$$

and (4.4.5) becomes

$$r\cos\theta_s = (z+z')\sin\theta_s \ . \tag{4.4.7}$$

The angle θ_s is defined in Fig. 4.4. The steepest descent evaluation of (4.4.1) is thus

$$I_1(\omega) \sim F(p_s) e^{\omega f(p_s)} e^{-i\pi/4} \left\{ \frac{2\pi}{\omega |f''(p_s)|} \right\}^{1/2} \ , \tag{4.4.8}$$

where

$$\begin{aligned} f''(p) &= -i(z+z')(\xi_1^{-1} + p^2 \xi_1^{-3}) \ , \\ &= -i(z+z')\xi_1^{-3} c_1^{-2} \ . \end{aligned} \tag{4.4.9}$$

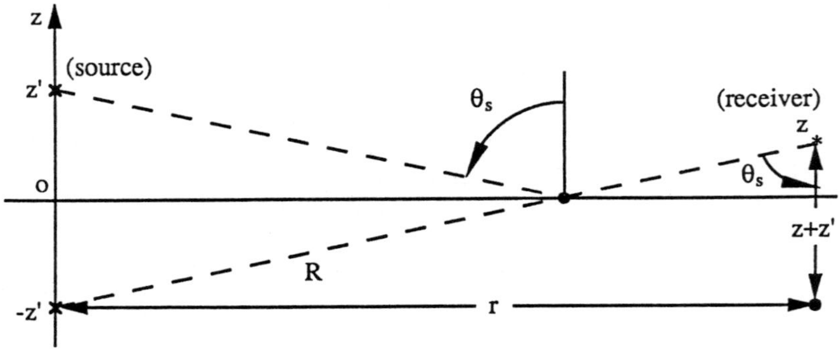

Fig. 4.4. Geometry for the steepest descent evaluation of the reflected field.

From the figure, we note that the result is that *the saddle point occurs at that value of p which yields the reflected ray between source and receiver.*

We also note that we can write

$$p_s = r\left[c_1 R\right]^{-1} \tag{4.4.10}$$

where

$$R = \left[r^2 + (z+z')^2\right]^{1/2} \ , \tag{4.4.11}$$

is the distance between the image source and the receiver. Using this we evaluate

$$f(p_s) = iR/c_1 \ , \tag{4.4.12}$$

and

$$f''(p_s) = -iR^2/\xi_{1s}(z+z') \ . \tag{4.4.13}$$

Combining these results in (4.4.8) we get the final asymptotic evaluation of (4.2.22) given by

$$\psi_r(r,z,z') \sim A\frac{e^{i\omega R/c_1}}{4\pi R}R(p_s) \ , \tag{4.4.14}$$

which has the form of a *spherical wave with amplitude proportional to the plane wave reflection coefficient evaluated at the saddle point (or specular point)* $R(p_s)$. We note that for large horizontal separation r between source and receiver we have $R \sim r$. Also the only place the frequency occurs is in the exponential, the phase term. It is possible to measure the result along the boundary and see the distortion of the wave shape due to $R(p_s)$. Measurement at two points will elicit the spherical spreading of the function.

We note finally that

$$p_s = c_1^{-1} \sin \theta_s \leq c_1^{-1} \quad , \tag{4.4.15}$$

so that the saddle point is located below c_1^{-1}. All along in this development we have assumed the method will work, i.e. that we actually can distort the contour through the saddle point. We next show that this is in fact possible by considering the only two cases we have, $c_1^{-1} < c_2^{-1}$ and $c_2^{-1} < c_1^{-1}$.

4.5 Case 1: $c_1 > c_2$

This is the case of a source in the fast medium (medium 1) and transmission into the slow medium (medium 2). From (4.4.15) assuming that $\theta_s \neq \pi/2$ we thus have

$$p_s < c_1^{-1} < c_2^{-1} \quad , \tag{4.5.1}$$

i.e. the saddle point is below both branch points. In order to determine whether the integration contour can be distorted into a contour of steepest descent we must first determine its general shape, and the shape of the region of steepest descent. The main barriers to the distortion are of course the branch cuts since they may be in the way of our attempts to distort the contour. For the moment, however, we neglect the branch cuts and focus on the distortion of the contour to the steepest descent path.

All we know about the integration contour up to now is that it runs from $-\infty$ to ∞ somehow and passes through the saddle point at an angle $-\pi/4$. The latter follows from the steepest descent discussion in the appendix. We next find the asymptotic regions of the steepest descent path. Recall that the integrand in (4.2.22) must decay along the latter and that steepest descent means fastest decay here. The asymptotic steepest descent paths are thus confined to the upper-half p-plane where $\operatorname{Im} p > 0$. We thus must find the asymptotes in the first and second quadrants

(a) $p \to \infty$ in the first quadrant: for this case we have that

$$\xi_1 = (c_1^{-2} - p^2)^{1/2} \sim +ip$$

where the + sign follows from either the square root definition in Sect. 4.3 or because $p_r > 0$ where $p = p_r + i\,p_i$. The exponential in (4.4.1) can thus be written as

$$e^{\omega f(p)} \sim e^{\omega\{-p_r(z+z')-p_i r+i(rp_r-p_i(z+z'))\}} \ .$$

Both real phase terms decay since $p_r > 0$, $p_i > 0$, $z + z' > 0$ and $r > 0$. On the steepest descent path we kill the oscillations. To do this simply set the imaginary part of the phase to zero, i.e.

$$p_i = \left[r/(z + z')\right] p_r$$

which is a straight line in the (p_r, p_i) plane with the same slope angle as that for the saddle point

$$\tan\theta_s = r/(z + z') \ ,$$

and intercept $p_i = 0$. This is illustrated below in Fig. 4.5.

(b) $p \to \infty$ in the second quadrant: For this case we have

$$\xi_1 = \left[c_1^{-2} - p^2\right]^{1/2} \sim -ip$$

which follows from either the square root definition in Sect. 4.3 or the fact that we require $\mathrm{Im}\,\xi_1 > 0$. The exponential in (4.4.1) is thus

$$e^{\omega f(p)} \sim e^{\omega\{(z+z')p_r-rp_i+i[rp_r+(z+z')p_i]\}} \ .$$

gain both phase terms decay ($p_r < 0$), and the condition to kill the oscillations is

$$p_i = -\left[r/(z + z')\right] p_r$$

which is a straight line with the negative of the slope in part (a). This is illustrated in Fig. 4.5.

Note that in Fig. 4.5 the steepest descent contour moves into the fourth quadrant as it passes through the saddle point. Asymptotically it ends in the first quadrant. It thus crosses the p_r axis and we find this point next. The steepest descent contour is defined by the phase term in (4.4.1),

$$rp + (z + z')\xi_1 \ .$$

this is an analytic function of p which equals R/c_1 at the saddle point $p_s = p_r = c_1^{-1}\sin\theta_s$ and $p_i = 0$. Thus we have the equation

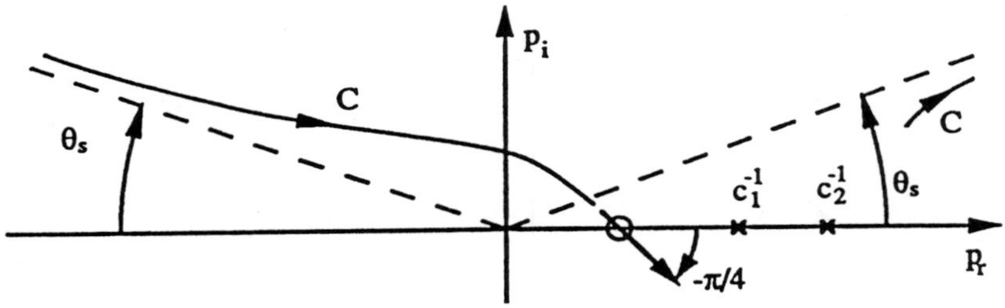

Fig. 4.5. Distortion of the contour C of integration in (4.2.22). The result is schematically illustrated as asymptotic to the steepest descent lines with slopes $\pm\tan\theta_s$ and passing through the saddle point p_s at angle $-\pi/4$. Branch points at c_1^{-1} and c_2^{-1} are also illustrated.

$$rp_r + (z+z')\left[c_1^{-2} - p_r^2\right]^{1/2} = R/c_1 \ . \tag{4.5.2}$$

To find a generalized form for this curve for complex p that reduces to (4.5.2) for $p_r = p_s$ we assume that it has the following form

$$rp + (z+z')\xi_1 = R/c_1 + ig^2(p) \ , \tag{4.5.3}$$

where the *rhs* is assumed to always have a positive imaginary part. Note that ξ_1 is real on part of the real axis ($p_i = 0, \quad |p_r| < c_1^{-1}$), but is not real in the complex plane even within a circle of radius c_1^{-1} centered at $p = 0$. To determine $g(p)$ we note that $g(p_s) = 0$. Also note that for $p_i = 0$ and $p_r \doteq p_0 > c_1^{-1}$ (4.5.3) becomes

$$rp_0 + i(z+z')\left[p_0^2 - c_1^{-2}\right]^{1/2} = R/c_1 + ig^2(p_0) \ . \tag{4.5.4}$$

Equating real and imaginary parts we have that

$$p_0 = R/rc_1 = \left[c_1 \sin\theta_s\right]^{-1} \ , \tag{4.5.5}$$

and

$$\begin{aligned} g^2(p_0) &= (z+z')\left[p_0^2 - c_1^{-2}\right]^{1/2} \\ &= (z+z')^2/rc_1 \ . \end{aligned} \tag{4.5.6}$$

The point $p_i = 0$, $p_r = p_0$ is the crossing point of our steepest descent curve. It is also greater than c_2^{-1} provided $\sin\theta_s < c_2/c_1$ and we assume

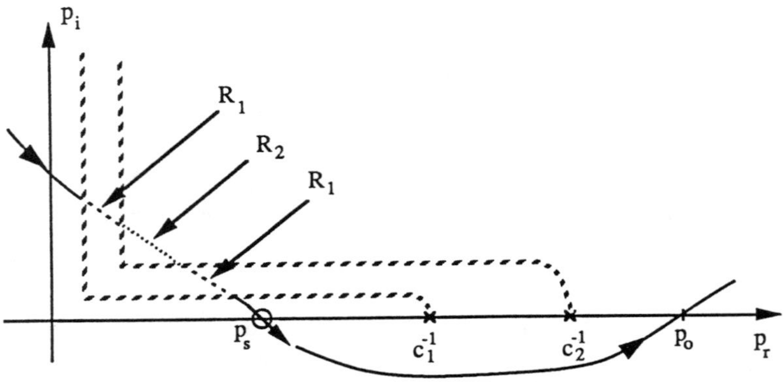

Fig. 4.6. Expanded view of the contour distortion in the first quadrant. Regions R_1 (dashed lines) are on the second sheet of the c_1^{-1} cut and region R_2 (dotted line) is on the second sheet of the c_2^{-1} cut. The point p_0 is the crossing point of the contour.

this. The crossing is illustrated below in Fig. 4.6. Since we know two real values of g we could for example parametrize it as a quadratic. Alternatively, we could find the curve numerically. Our purpose above has not been to completely specify the curve but rather to concentrate on its general shape with only some specific points. This is sufficient for our purpose.

We have thus distorted the integration contour along the lines of steepest descent. Now we must consider whether we *can* in fact do this because of the presence of branch cuts in the first quadrant. Clearly, since we showed that there are no singularities in the second or fourth quadrants, the contour distortion can be done in these quadrants. An expanded view of the contour in the first quadrant is illustrated in Fig. 4.6.

Referring to Fig. 4.6 we note that the contour crosses both branch cuts twice so that when it reaches the saddle point it is again on the correct physical sheet. The regions R_1 and R_2 are on the steepest ascent path toward p_s so that the integral is still dominated by the saddle point contribution. The region R_2 is on a sheet where $\xi_{2i} < 0$ and this is not important in the reflected field in (4.4.1) since ξ_2 does not occur in the phase. However, once we cross the c_1^{-1} cut the sheet has $\xi_{1i} < 0$ and we must minimize this travel path. We assume we can do this either by minimizing the actual path on this sheet or by distorting the steepest descent path around the branch cut so that it

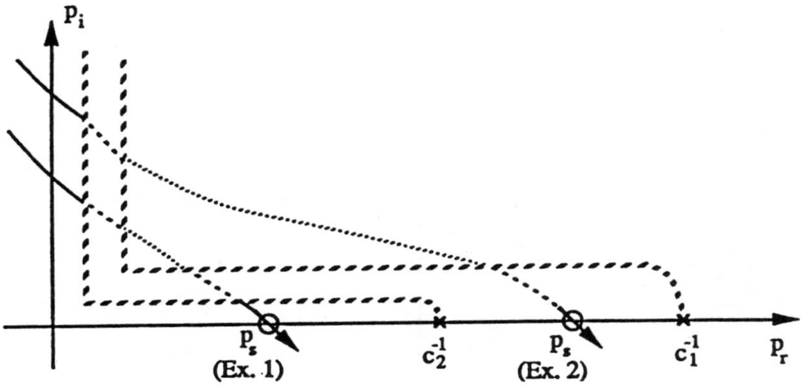

Fig. 4.7. Two examples for the saddle point (p_s) position for the case when $c_1 < c_2$. For Example 1 the steepest descent contour passes through the saddle point on the correct sheet. For Example 2 the naive steepest descent contour appears to pass through the saddle point but is on the wrong sheet. The details on the figure are considerably expanded for clarity.

remains on the first sheet. This may appear to be a bit of sleight-of-hand but it must be remembered that the steepest descent method is only an approximation anyway. Its true test here is how well it represents the experimental results.

The final result for $c_1 > c_2$ and $\sin\theta_s < c_2/c_1$ is that to leading order the reflected field (4.4.1) is approximately given by the result

$$\psi_r(r,z,z') \sim A\frac{e^{i\omega R/c_1}}{4\pi R}R(p_s) \ , \tag{4.5.7}$$

as discussed in (4.4.14). To this order no other terms must be considered. This will not be true always for Case 2 ($c_1 < c_2$) in the next section. There we will find that the contour may naturally wrap around a branch cut giving rise to an additional term in the asymptotic expansion.

4.6 Case 2: $c_1 < c_2$

This is the case of a source in the slow medium (1) with transmission to the faster medium (2). Since the phase term in (4.2.22) doesn't involve ξ_2 the same saddle point at $p_s = c_1^{-1}\sin\theta_s$ occurs here. The

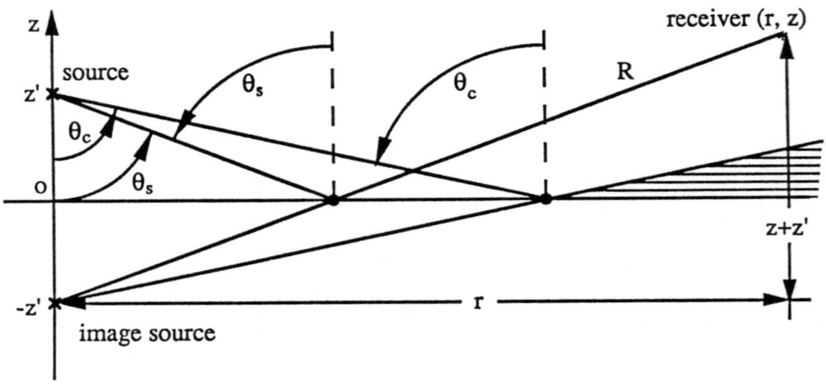

Fig. 4.8. Example 1, $c_1 < c_2$, $\theta_s < \theta_c$ in the text. The receiver sees only a spherical wave which appears to emanate from the image source.

position of the saddle point thus depends on the sound speed in the source region but is independent of the sound speed of the second medium. Unlike Case 1, we can have two different examples for this case.

Example 1 is when $p_s < c_2^{-1}$ and Example 2 when $c_2^{-1} < p_s < c_1^{-1}$. These are illustrated in Fig. 4.7 where we also illustrate that Example 1 has the same properties as Case 1 but Example 2 does not. For the latter the naive sketch of the steepest descent path appears to pass through the saddle point but is on the wrong sheet. This example is different and will require additional consideration.

First, recall from Snell's law in Chap. 3 that the critical angle is found when the angle of the wave in region 2 is $\pi/2$, i.e.

$$\theta_c = \sin^{-1}(c_1/c_2) \ .$$

Example 1:

$$\begin{aligned} p_s &< c_2^{-1} \ , \\ \sin\theta_s &< (c_1/c_2) = \sin\theta_c \ , \\ \theta_s &< \theta_c \ . \end{aligned} \tag{4.6.1}$$

Example 2:

$$\begin{aligned} p_s &= c_1^{-1}\sin\theta_s > c_2^{-1} \ , \\ \sin\theta_s &> (c_1/c_2) = \sin\theta_c \ , \\ \theta_s &> \theta_c \ . \end{aligned} \tag{4.6.2}$$

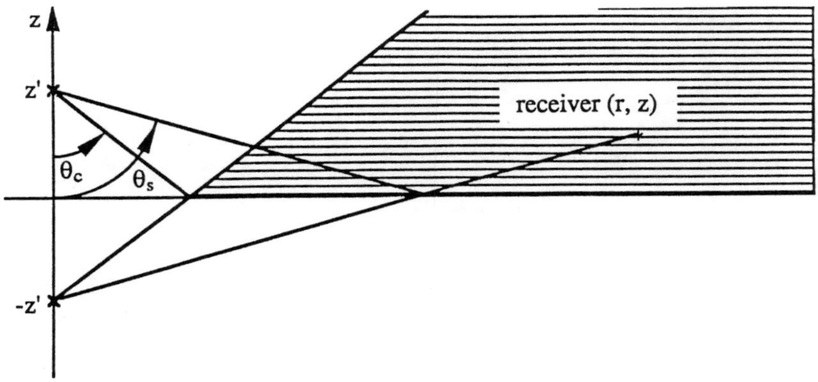

Fig. 4.9. Example 2, $c_1 < c_2$, $\theta_s > \theta_c$ in the text. Scattering occurs at greater than the critical angle and the receiver views a wide angle reflected spherical wave plus a head wave.

Reflected rays for these two examples are illustrated in Figs. 4.8 and 4.9. For Example 1 in Fig. 4.8, the asymptotic (ray) development illustrates that the receiver sees only the real reflected ray. For Example 2 in Fig. 4.9 it will be shown below using asymptotic analysis and a distortion of the steepest descent contour that the receiver (now in the shadow region) will see a reflected wave which has a phase modification (the so-called wide angle reflected ray due to total internal reflection) and, in addition, a wave with different asymptotic properties (in both frequency and range) that travels along the boundary and is called a head wave.

For Example 1 we thus have no problem with the distortion of the steepest descent contour and the result for the reflected field is the same as that in Sect. 4.5, viz.

$$\psi_r(r,z,z') \sim A\frac{e^{i\omega R/c_1}}{4\pi R}R(p_s) \ , \tag{4.6.3}$$

i.e. a spherical wave modified by the plane wave reflection coefficient evaluated at the saddle point.

For Example 2 we note from Fig. 4.7 that we crossed the branch cut at c_1^{-1} twice thus returning to its correct sheet. The problem is the branch cut at c_2^{-1} which has only been crossed once so that the path through the saddle point is on the wrong sheet of this branch cut. We must define an alternative steepest descent path to either (a)

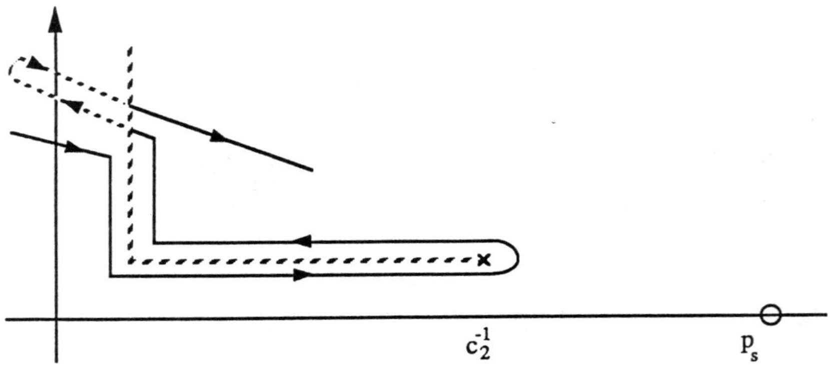

Fig. 4.10. Distortion of the steepest descent contour about the branch cut starting at c_2^{-1} (offset for clarity). The dashed contours on the second sheet cancel each other.

avoid crossing this branch cut at all or (b) cross it twice. The latter is possible and is illustrated in Fig. 4.10. The new contour crosses the branch cut twice. It then goes through the branch cut from c_1^{-1} twice (as in Case 1), and then goes through the saddle point. There are now two contributions. The first is the saddle point but now with $p_s > c_2^{-1}$ so that the reflection coefficient has unit magnitude and is only modified in phase. From (4.2.15) and our square root definition

$$\xi_2 = (c_2^{-2} - p_2)^{1/2} = i(p^2 - c_2^{-2})^{1/2}, \qquad p > c_2^{-1} \ ,$$

we have that for $p = p_s > c_2^{-1}$

$$\begin{aligned} R(p_s) &= \frac{\rho_2(c_1^{-2} - p_s^2)^{1/2} - i\rho_1(p_s^2 - c_2^{-2})^{1/2}}{\rho_2(c_1^{-2} - p_s^2)^{1/2} + i\rho_1(p_s^2 - c_2^{-2})^{1/2}} \\ &\doteq \exp(-2i\Phi_s) \ , \end{aligned} \tag{4.6.4}$$

where

$$\tan \Phi_s = \rho_1 |\xi_{2s}| / \rho_2 \xi_{1s} \ . \tag{4.6.5}$$

The saddle point contribution thus consists solely of a phase modified spherical wave

$$\psi_s(r, z, z') \sim A \frac{e^{i\omega R/c_1}}{4\pi R} e^{-2i\Phi_s} \ . \tag{4.6.6}$$

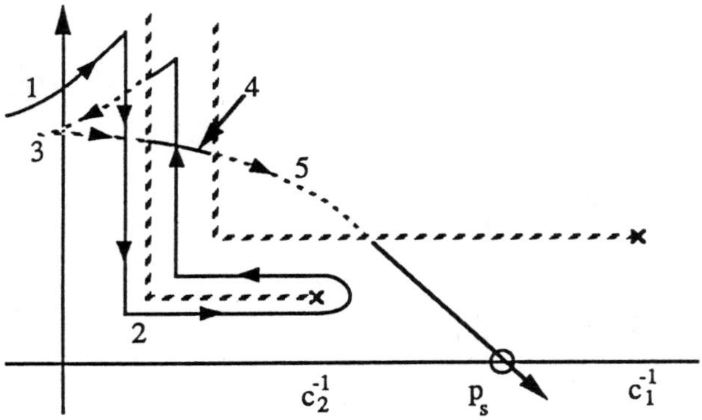

Fig. 4.11. Full contour distortion around the c_2^{-1} branch cut and the resulting steepest descent contour. The original contour (1) can be distorted in the second quadrant (no singularities). The resulting contour wraps around the branch cut (2), crosses back on the second sheet of this cut (3) and then resumes our original steepest descent path coming back onto the correct sheet (4) and crossing the c_1^{-1} branch cut twice (5) to intercept the saddle point on the physical sheet.

Since $\theta_s > \theta_c$ for this case, the result (4.6.6) is also termed a *wide-angle reflection.* The second contribution is a line integral around the branch cut starting at c_2^{-1}. This cannot be avoided here since we are required to distort the contour around this cut. The result is illustrated in Fig. 4.11. The resulting branch line integral (BLI) contribution to ψ can be written using (4.2.22) as the sum of four integrals

$$\psi_{BLI}(r,z,z') \sim \frac{A}{8\pi}\left(\frac{2i\omega}{\pi r}\right)^{1/2}\left\{\int_{i\infty}^{0}(L) + \int_{0}^{c_2^{-1}}(L) + \int_{c_2^{-1}}^{0}(U) + \int_{0}^{i\infty}(R)\right\} , \qquad (4.6.7)$$

evaluated along the left or lower (L) lip of the cut or along the upper (U) or right (R) lip. The integrals can be written explicitly using (4.2.22) and (4.4.2) as

$$\int_{i\infty}^{0}(L) = \int_{i\infty}^{0} R_-(p)p^{1/2}\xi_1^{-1}\exp\big[\omega f(p)\big]\,dp \ , \qquad (4.6.8)$$

where

$$R_-(p) = R(p)\Big|_{\alpha_2=-\pi/2;\, \beta_2=\pi/2} . \tag{4.6.9}$$

The second integral is

$$\int_0^{c_2^{-1}} (L) = \int_0^{c_2^{-1}} R_L(p) p^{1/2} \xi_1^{-1} \exp\Big[\omega f(p)\Big] dp , \tag{4.6.10}$$

with

$$R_L(p) = R(p)\Big|_{\alpha_2=\beta_2=0} . \tag{4.6.11}$$

The third integral is

$$\int_{c_2^{-1}}^{2} (U) = \int_{c_2^{-1}}^{0} R_U(p) p^{1/2} \xi_1^{-1} \exp\Big[\omega f(p)\Big] dp , \tag{4.6.12}$$

with

$$R_U(p) = R(p)\Big|_{\alpha_2=2\pi;\, \beta_2=0} , \tag{4.6.13}$$

and the last integral is

$$\int_0^{i\infty} (R) = \int_0^{i\infty} R_R(p) p^{1/2}\, \xi_1^{-1} \exp\Big[\omega f(p)\Big] dp , \tag{4.6.14}$$

with

$$R_R(p) = R(p)\Big|_{\alpha_2=3\pi/2;\, \beta_2=\pi/2} . \tag{4.6.15}$$

The angle definitions of the ξ_2-branch cut were illustrated in Fig. 4.3. We define the discontinuity of the reflection coefficient as

$$\Delta R(p) = R(p)\Big|_{\alpha_2+\beta_2=0} - R(p)\Big|_{\alpha_2+\beta_2=2\pi} \tag{4.6.16}$$

$$= \frac{\rho_2\xi_1 - \rho_1|\xi_2|}{\rho_2\xi_1 + \rho_1|\xi_2|} - \frac{\rho_2\xi_1 + \rho_1|\xi_2|}{\rho_2\xi_1 - \rho_1|\xi_2|}$$

$$= \frac{-4\rho_1\rho_2\xi_1\xi_2}{(\rho_2\xi_1)^2 - (\rho_1\xi_2)^2} , \tag{4.6.17}$$

where the absolute value on ξ_2 is unnecessary since ξ_2 is purely real for $p_i = 0$, $0 < p_r < c_2^{-1}$ (real axis) and $p_r = 0$ (positive imaginary axis). Using this we can write the four integrals in (4.6.7) as

$$BLI = -\int_{c_2^{-1}}^{i\infty} \Delta R(p) p^{1/2}\, \xi_1^{-1} \exp\Big[\omega f(p)\Big] dp , \tag{4.6.18}$$

where the integral runs along the real axis from c_2^{-1} to 0, then along the imaginary axis. Note that $f(p)$ is a regular analytic function of p along this path. The full contribution to the reflected field is thus from (4.6.7) and (4.6.18)

$$\psi_{BLI}(r,z,z') \sim \frac{A}{8\pi}\left(\frac{2i\omega}{\pi r}\right)^{1/2} BLI \ , \tag{4.6.19}$$

and we must evaluate (4.6.18). We do this below.

Recall from (4.4.2) that

$$f(p) = i\Big[rp + (z+z')\xi_1\Big] \ .$$

Along that part of the BLI contour on the imaginary axis $p = ip_i$, and $p_i > 0$ so that $f(p)$ decays in range and oscillates in the depth variables. Along the contour $0 < p < c_2^{-1}$, $f(p)$ is pure imaginary and thus $\exp[\omega f(p)]$ oscillates in both range and depth. In the steepest descent evaluation our purpose was to minimize the oscillations of $f(p)$ and we were in fact able to set $f'(p) = 0$. Here we are unable to do the latter, but we are able to minimize $f'(p)$, i.e. the main contributions to the integral comes from those p-values where Im $f'(p)$ is a minimum or from those p-values representing slower oscillations. We have that

$$f'(p) = i\left[r - (z+z')p\xi_1^{-1}\right] ,$$

and along the contour $0 < p < c_2^{-1}$

$$\text{Im}\ f'(p) = r - (z+z')p\,\xi_1^{-1} \ .$$

This is a minimum when ξ_1 is a minimum and the latter occurs at the branch point $p = c_2^{-1}$. For example we have that

$$\xi_1\Big|_{p=0} = c_1^{-1} \ ; \quad \text{Im}\ f'(0) = r \ ,$$

and

$$\xi_1\Big|_{c_2^{-1}} = \left[c_1^{-2} - c_2^{-2}\right]^{1/2} < c_1^{-1} \ ,$$

where

$$\text{Im}\ f'(c_2^{-1}) = r - \frac{z+z'}{c_2(c_1^{-2} - c_2^{-2})^{1/2} < r} \ . \tag{4.6.20}$$

Along the vertical contour, $p = ip_i$, and

$$f(p) = i\left[ir\,p_i + (z+z')\left[c_1^{-2}+p_i^2\right]^{1/2}\right]$$

so that

$$\mathrm{Re}\ f(p) = -rp_i$$

which gives decay in range and

$$\mathrm{Im}\ f(p) = (z+z')\left[c_1^{-2}+p_i^2\right]^{1/2}$$

which is larger than on the horizontal contour and

$$\begin{aligned}\mathrm{Im}\, f'(p) &= \mathrm{Im}\left\{i\left[r - i(z+z')p_i\left[c_1^{-2}+p_i^2\right]^{-1/2}\right]\right\}\\ &= r\end{aligned}$$

The evaluation of (4.6.18) thus consists in expanding the integrand about the branch point c_2^{-1}, the point of slowest oscillation of the integral.

To evaluate (4.6.18) we first expand the phase about the branch point c_2^{-1}

$$f(p) = f(c_2^{-1}) + f'(c_2^{-1})(p - c_2^{-1}) + \cdots. \tag{4.6.21}$$

We can do this expansion because f is an analytic function in the neighborhood of c_2^{-1}. Note also that unlike the steepest descent evaluation where we had to go to quadratic terms since $f'(p_s) = 0$, here we stop the expansion at the first order or derivative term. In addition we expand the amplitude in (4.6.18) to zeroth order. Everything in the amplitude is analytic near c_2^{-1} except for one of the square roots in the ξ_2 term in $\Delta R(p)$. The latter remains inside the integral while everything else comes outside. The result can be written as

$$BLI \sim B \exp\left[\omega f(c_2^{-1})\right] J\ , \tag{4.6.22}$$

where

$$\begin{aligned}B &\doteq \left\{\frac{4\rho_1\rho_2 p^{1/2}\left[c_2^{-1}+p\right]^{1/2}}{(\rho_2\xi_1)^2 - (\rho_1\xi_2)^2}\right\}_{p=c_2^{-1}}\ ,\\ &= \frac{4\cdot 2^{1/2}\rho_1 c_2}{\rho_2\left[(c_2/c_1)^2 - 1\right]}\ ,\end{aligned} \tag{4.6.23}$$

and the remaining integral J is given by

$$J = \int_{c_2^{-1}}^{i\infty} (c_2^{-1} - p)^{1/2} \exp\left\{\omega f'(c_2^{-1})(p - c_2^{-1})\right\} dp \ . \tag{4.6.24}$$

To evaluate J, shift the integration to c_2^{-1} and rotate the contour, i.e. introduce the transformation

$$p - c_2^{-1} = iy^2 \ . \tag{4.6.25}$$

Since we have that $f'(c_2^{-1}) = i\,\mathrm{Im}\left[f'(c_2^{-1})\right]$ and $\mathrm{Im}\ f'(c_2^{-1}) > 0$ for large enough r by (4.6.20) we can write J as

$$J = 2e^{i\pi/4} \int_0^\infty y^2 e^{-ay^2} dy \ , \tag{4.6.26}$$

where

$$a = \omega |f'(c_2^{-1})| \ . \tag{4.6.27}$$

The latter integral is just

$$\int_0^\infty y^2 e^{-ay^2} dy = \frac{1}{4} \frac{\pi^{1/2}}{a^{3/2}} \ . \tag{4.6.28}$$

Using the fact that

$$f(c_2^{-1}) = i\left\{rc_2^{-1} + (z + z')(c_1^{-2} - c_2^{-2})^{1/2}\right\} \tag{4.6.29}$$

along with (4.6.23) we can thus write (4.6.22) as

$$BLI \sim \frac{2(2\pi i)^{1/2} c_2}{(c_2/c_1)^2 - 1} \frac{\rho_1}{\rho_2} \frac{e^{i\omega t_h}}{\omega^{3/2} L^{3/2}} \tag{4.6.30}$$

where

$$t_h \doteq \frac{r}{c_2} + (z + z')(c_1^{-2} - c_2^{-2})^{1/2} \ , \tag{4.6.31}$$

and

$$L \doteq |f'(c_2^{-1})| = r - \frac{(z + z')}{c_2(c_1^{-2} - c_2^{-2})^{1/2}} \ . \tag{4.6.32}$$

These latter terms have a geometric interpretation using the critical angle definition $\sin\theta_c = c_1/c_2$. We note that L becomes

$$L = r - (z + z')\tan\theta_c \ , \tag{4.6.33}$$

and t_h can be written as

$$t_h = \frac{r}{c_2} + \frac{\cos\theta_c}{c_1}(z + z') \ . \tag{4.6.34}$$

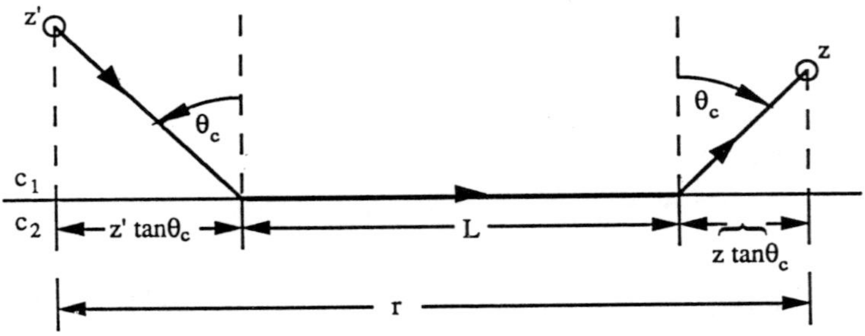

Fig. 4.12. Propagation of a lateral or headwave due to total internal reflection. The wave propagates in region 1 a distance $(z+z')\sec\theta_c$ at speed c_1, and laterally a distance L with speed c_2 in region 2.

Substituting for r from (4.6.33) we get that

$$t_h = \frac{L}{c_2} + \frac{\sec\theta_c}{c_1}(z+z') \ . \tag{4.6.35}$$

The latter is interpreted as the travel time for a wave from source point z' striking the boundary at the critical angle, propagating along the boundary a distance L with the wave speed of the lower medium c_2, then radiating at the critical angle to the receiver point z. The distance travelled in region 1 is $(z+z')\sec\theta_c$ at wave speed c_1. The interpretation of the wave is due to its lateral propagation and it is referred to as a *lateral wave* or a *head wave*. See Fig. 4.12.

The total reflected field for Example 2 can thus be written by combining (4.6.6), (4.6.19) and (4.6.30) as

$$\psi_r(r,z,z') \sim \psi_s(r,z,z') + \psi_{BLI}(r,z,z') \ , \tag{4.6.36}$$

so that

$$\psi_r(r,z,z') \sim A\frac{e^{i\omega R/c_1}}{4\pi R}e^{-2i\Phi_s} + \frac{iA}{2\pi}\frac{\rho_1 c_2}{\rho_2\left[(c_2/c_1)^2-1\right]}\frac{e^{i\omega t_h}}{\omega r^{1/2}L^{3/2}} \ . \tag{4.6.37}$$

The two terms in (4.6.37) have different decay properties. The saddle point term spreads spherically (R^{-1}) in space and its amplitude is independent of frequency. For long range ($L \sim r$) the branch line integral term decays spatially like r^{-2} and its amplitude falls inversely in frequency (ω^{-1}). So the branch line integral term decays faster in both range and frequency. Despite this fact the two terms are inextricably coupled by our contour distortion and asymptotic analysis.

In addition, for large range we have from (4.6.34) that

$$t_h \sim r/c_2 \ .$$

For the saddle point contribution (reflected ray) the traveltime is $t_s = R/c_1$ which for large range is

$$t_s \sim r/c_1 \ ,$$

which implies that for large range $t_h < t_s$, i.e. the *head wave arrives before the reflected wave.* In addition the ω^{-1} behavior of the head wave corresponds in Fourier transform space to an integration in time. This implies that the head wave has a smoother wave form than the saddle contribution and that it has a longer tail (because of the temporal integration) and a faster rise. This corresponds to what is actually observed in the temporal signal.

4.7 Transmitted Field

We write the transmitted field from (4.2.23) as

$$\psi_t(r,z,z') \sim \frac{A}{8\pi}\left(\frac{2i\omega}{\pi r}\right)^{1/2} I_2(\omega) \ , \tag{4.7.1}$$

where

$$I_2(\omega) = \int_{-\infty}^{\infty} T(p)p^{1/2}\xi_1^{-1}e^{\omega g(p)}dp \ , \tag{4.7.2}$$

with

$$g(p) = i\big[pr + \xi_1 z' + \xi_2|z|\big] \ . \tag{4.7.3}$$

Since z is negative in region 2, we have written an absolute value in (4.7.3). The factor ξ_1^{-1} in the amplitude of the integrand of (4.7.2) actually cancels a similar factor in the transmission coefficient (4.2.16). A major difference from the reflected field occurs in the phase term (4.7.3) in that both square roots are now present. In addition, we find from (4.7.3) that

$$g'(p) = i(r - p\xi_1^{-1}z' - p\xi_2^{-1}|z|) \ , \tag{4.7.4}$$

and

$$g''(p) = -i(c_1^{-2}\xi_1^{-3}z' + c_2^{-2}\xi_2^{-3}|z|) \ . \tag{4.7.5}$$

The saddle point p_s occurs when $g'(p_s) = 0$ or for

$$r = p_s \xi_{1s}^{-1} z' + p_s \xi_{2s}^{-1} |z| \ . \tag{4.7.6}$$

If we write $g = u + iv$, the steepest descent evaluation of (4.7.2) yields, using Appendix 4A,

$$I_2(\omega) \sim p_s^{1/2} \xi_{1s}^{-1} T(p_s) e^{\omega g(p_s)} e^{i\pi \operatorname{sgn}(v''(p_s))/4}$$

$$\cdot \left\{ 2\pi |v''(p_s)| / \omega |g''(p_s)| \right\}^{1/2} \ , \tag{4.7.7}$$

where we must explicitly evaluate the various terms. We consider different cases and examples as we did for the reflected field.

For Case 1 ($c_1 > c_2$ and $p_s < c_1^{-1} < c_2^{-1}$) and Case 2, Example 1 ($c_2 > c_1$ and $p_s < c_2^{-1} < c_1^{-1}$) both square roots in the phase are real. We parametrize them as

$$\xi_{1s} = (c_1^{-2} - p_s^2)^{1/2} = c_1^{-1} \cos\theta_s \ , \tag{4.7.8}$$

and

$$\xi_{2s} = (c_2^{-2} - p_s^2)^{1/2} = c_2^{-1} \cos\theta_t \ , \tag{4.7.9}$$

where by Snell's law the saddle position is given by

$$p_s = c_1^{-1} \sin\theta_s = c_2^{-1} \sin\theta_t \ . \tag{4.7.10}$$

The geometry is defined in Fig. 4.13. Using these results we can write (4.7.6) as

$$r = \alpha_1 + \alpha_2; \quad \alpha_1 = z' \tan\theta_s; \quad \alpha_2 = |z| \tan\theta_t \tag{4.7.11}$$

and

$$\begin{aligned} g(p_s) &= i(p_s r + \xi_{1s} z' + \xi_{2s} |z|) \ , \\ &= i(R_1 c_1^{-1} + R_2 c_2^{-1}) \ ; \end{aligned} \tag{4.7.12}$$

where $R_1 = z' \sec\theta_s$ and $R_2 = |z| \sec\theta_t$ are the radial distances from the source to the boundary and from the boundary to the field point z respectively. They are illustrated in Fig. 4.13. To evaluate (4.7.12) we substituted for r from (4.7.11) and p_s from (4.7.10). In addition, from (4.7.5) we have that

$$\begin{aligned} g''(p_s) &= -i(c_1^{-2} \xi_{1s}^{-3} z' + c_2^{-2} \xi_{2s}^{-3} |z|) \ , \\ &= iv''(p_s) \ , \end{aligned} \tag{4.7.13}$$

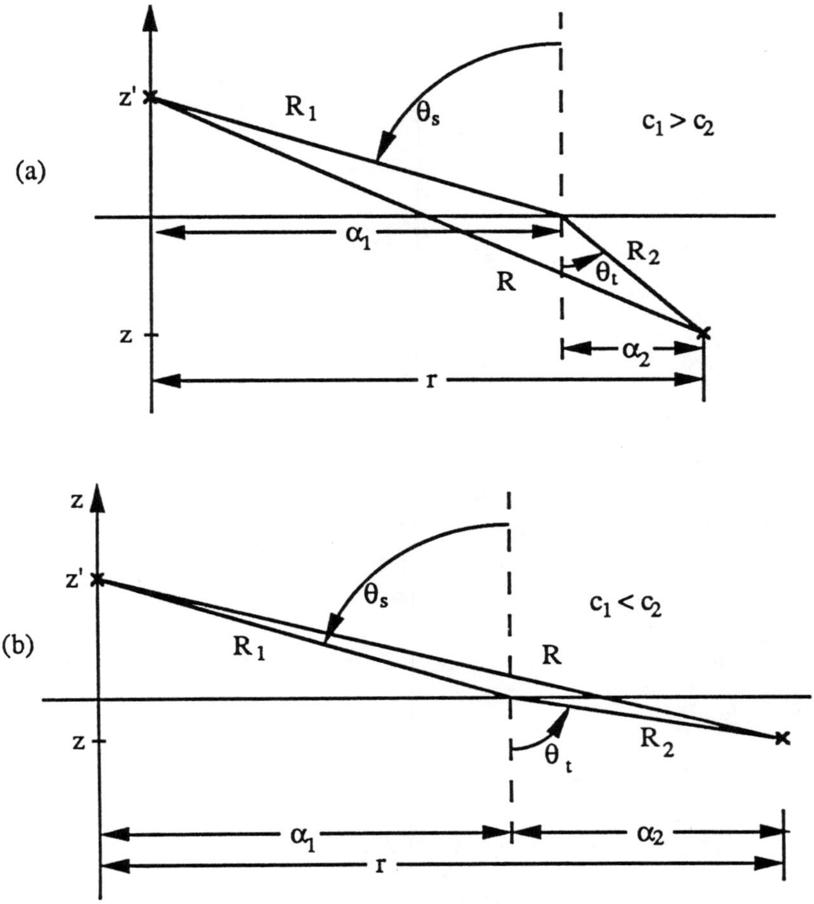

Fig. 4.13. Steepest descent (ray) paths for a source at z' (medium 1) and receiver at z (medium 2) for the transmitted field. In (a) the ray bends towards the normal and in (b) towards the horizontal.

so that

$$|v''(p_s)| = |g''(p_s)|$$

$$= c_1 R_1 \sec^2\theta_s + c_2 R_2 \sec^2\theta_t \ . \tag{4.7.14}$$

Combining these results in (4.7.7) and the latter in (4.7.1) we can write

$$\psi_t(r,z,z') \sim \frac{AT(p_s)}{4\pi(r_1 r_2)^{1/2}} e^{i\omega(R_1/c_1 + R_2/c_2)} \ , \tag{4.7.15}$$

where we have defined

$$r_1 = R_1 + R_2(c_2/c_1) \ , \tag{4.7.16}$$

and

$$r_2 = R_1 + R_2(c_2\cos^2\theta_s / c_1\cos^2\theta_t) \ . \tag{4.7.17}$$

For long range the transmitted field thus behaves like a spherical wave distorted by the change in material parameters. For $c_2 = c_1(\theta_s = \theta_t)$ the exact spherical wave nature is apparent. Travel times $R_1 c_1^{-1}$ and $R_2 c_2^{-1}$ in each medium are obvious. Note that in each medium separately the wave appears almost as a cylindrical wave, and the overall result the product of two almost cylindrical waves. This is necessary in the amplitude factors for just a single spherical wave to emerge in the case $c_2 = c_1$.

For Case 2 ($c_2 > c_1$ and $c_2^{-1} < p_s < c_1^{-1}$) we have that ξ_{2s} is pure imaginary

$$\xi_{2s} = i(p_s^2 - c_2^{-2})^{1/2}$$

and from (4.7.6) the saddle point would occur at

$$r = z'\tan\theta_s - i|z|\tan\theta_t \ .$$

For the receiver off the boundary ($z \neq 0$) the only solution can occur if $\theta_t = 0$. To preserve Snell's law we must conclude that $\theta_s = 0$ so that $r = 0$, i.e. no propagation in range is possible. Further, $p_s = 0$, and this contradicts our case assumption. This is not interesting. Our conclusion is that no real saddle point exists. What remains are integrals around the branch cuts illustrated in Fig. 4.14. The asymptotic values of the overall contour are the same as those illustrated in Fig. 4.5. The transmitted field can thus be written as

$$\psi_t(r,z,z') \sim \frac{A}{8\pi}\left(\frac{2i\omega}{\pi r}\right)^{1/2}\left\{BLI(1) + BLI(2)\right\} \ , \tag{4.7.18}$$

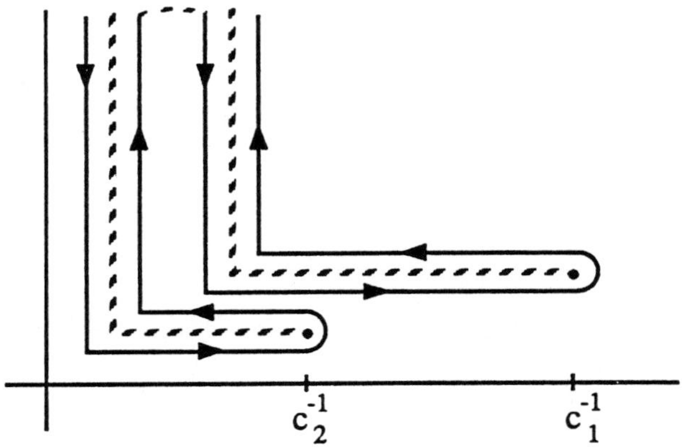

Fig. 4.14. Branch line integrals for the transmitted field for the case where no saddle point is present.

where the branch line integrals $BLI(j)$ are written as contour integrals $(j = 1, 2)$ using the contours illustrated in Fig. 4.14

$$BLI(j) = 2\rho_1 \int_{C_j} \frac{p^{1/2} e^{\omega g(p)}}{\rho_2 \xi_1 + \rho_1 \xi_2} dp \ . \tag{4.7.19}$$

Here we have explicitly substituted for the transmission coefficient from (4.2.16). The C_j contours surround branch cuts starting at c_j^{-1}. These are a different type of branch line integral than the one which occurred in Sect. 4.6. The latter contained a branch point only in the amplitude of the integrand whereas here we have a branch point in both the amplitude and phase for both integrals. We rewrite (4.7.19) as the sum of two integrals

$$BLI(j) = \rho_2 U(j) - \rho_1 V(j) \ , \tag{4.7.20}$$

where

$$U(j) = \int_{C_j} \xi_1 E(p) e^{\omega g(p)} dp \ , \tag{4.7.21}$$

and

$$V(j) = \int_{C_j} \xi_2 E(p) e^{\omega g(p)} dp \ , \tag{4.7.22}$$

where

$$E(p) = 2\rho_1 p^{1/2} (\rho_2^2 \xi_1^2 - \rho_1^2 \xi_2^2)^{-1} \ , \tag{4.7.23}$$

is analytic at both branch points c_1^{-1} and c_2^{-1}.

To evaluate $BLI(1)$, expand the analytic parts of the amplitude of the integrands and the analytic part of the phase

$$g_1(p) = i(pr + \xi_2|z|) \tag{4.7.24}$$

in a truncated Taylor expansion about c_1^{-1}. We can write

$$U(1) \approx (2/c_1)^{1/2} f_1 \exp(\omega g_{11}) \int_{C_1} (c_1^{-1} - p)^{1/2} \exp\Big\{\omega g_{11'}(p - c_1^{-1}) + (2/c_1)^{1/2} z'(c_1^{-1} - p)^{1/2}\Big\} dp \ , \tag{4.7.25}$$

where $f_1 = E(c_1^{-1})$, $g_{11} = g_1(c_1^{-1})$, and $g'_{11} = g'_1(c_1^{-1})$. Because of analyticity the contour integral can be written as a vertical contour integral above the branch point c_1^{-1}. (We treat this in more detail in Chap. 5. See Fig. 5.10). Further, under the transformation

$$c_1^{-1} - p = -ix \ , \tag{4.7.26}$$

which rotates and translates the contour so that it becomes a loop integral centered at $x = 0$ along the real axis, the result can be written as

$$U(1) \approx U_1 \int_{C_+} x^{1/2} \exp(-a_1 x + b_1 x^{1/2}) dx \ , \tag{4.7.27}$$

where

$$a_1 = -i\omega g'_{11} \ , \tag{4.7.28}$$

$$b_1 = \omega z'(2/ic_1)^{1/2} \ , \tag{4.7.29}$$

C_+ defines the loop integral contour, and

$$U_1 = (2i/c_1)^{1/2} f_1 \exp(\omega g_{11}) \ . \tag{4.7.30}$$

Evaluating the loop integral to make it a real integral and introducing the further transformation

$$a_1 x = u^2 \tag{4.7.31}$$

yields the result

$$U(1) \approx -2U_1 a_1^{-3/2} \int_0^\infty u^2 e^{-u^2} (e^{-d_1 u} + e^{d_1 u}) du \ , \tag{4.7.32}$$

where

$$d_1 = b_1 a_1^{-1/2} \ . \tag{4.7.33}$$

For r large we note that

$$a_1 \sim \omega r, \ d_1 \sim O(r^{-1/2}) \ . \tag{4.7.34}$$

Expanding the integrand in (4.7.32) to lowest order and evaluating the resulting integral we get, after some algebra, that

$$U(1) \approx -2(2\pi i)^{1/2}(c_2^2/\rho_1 c_1 K^2)(\omega r)^{-3/2} \cdot \exp\Big[i\omega r/c_1 - \omega|z|K/c_2\Big] \ , \tag{4.7.35}$$

where

$$K = \left\{(c_2/c_1)^2 - 1\right\}^{1/2} \ . \tag{4.7.36}$$

Thus $U(1)$ consists of a horizontal wave in r plus a decaying wave modified by the second medium and receiver distance from the boundary. The wave also decays algebraically in frequency and range as $(\omega r)^{-3/2}$.

Next, expand $V(1)$ from (4.7.22) about the branch point c_1^{-1}. We note that the amplitude of the integrand is analytic at this point and the analytic part of the phase is the same as that for $U(1)$. The result is

$$V(1) \approx F_1 \exp(\omega g_{11}) \int_{C_1} \exp\Big\{\omega g'_{11}(p - c_1^{-1}) + (2/c_1)^{1/2} z'(c_1^{-1} - p)^{1/2}\Big\} dp \ , \tag{4.7.37}$$

where

$$F_1 = \Big\{\xi_2 E(p)\Big\}_{c_1^{-1}} \ , \tag{4.7.38}$$

and g_{11} and g'_{11} are defined following (4.7.25). Use the transformation (4.7.26), evaluate the resulting loop integral and use the further transformation (4.7.31) to yield the real integral

$$V(1) \approx 2iF_1 \exp(\omega g_{11}) a_1^{-1} \int_0^\infty u\, e^{-u^2}(e^{-d_1 u} - e^{d_1 u}) du \ , \tag{4.7.39}$$

where d_1 is given by (4.7.33). For r large we again expand the integrand in (4.7.39) (to lowest order yielding $-2d_1 u$) and evaluate the result to yield

$$V(1) \approx -2i(2\pi i)^{1/2}(c_2 z'/\rho_1 c_1 K)\omega^{-1/2} r^{-3/2} \cdot \exp\Big[i\omega r/c_1 - \omega|z|K/c_2\Big] \ , \tag{4.7.40}$$

which has the same wave shape as $U(1)$ but decays algebraically slower in frequency ($\omega^{-1/2}$).

Combining the results using (4.7.20) we can write that

$$BLI(1) \sim B_1 \omega^{-1/2} r^{-3/2} \exp\Big\{ i\omega r/c_1 - \omega|z|K/c_2 \Big\} \ , \tag{4.7.41}$$

where

$$B_1 = -2(2\pi i)^{1/2}(c_2/c_1K)\Big([\rho_2 c_2/\rho_1 \omega K] - iz'\Big) \ . \tag{4.7.42}$$

Finally note that the decaying exponential in (4.7.41) results from expanding about the higher branch point ($c_1^{-1} > c_2^{-1}$).

Analogous results follow for $U(2)$ and $V(2)$. In this instance we expand around the branch point at c_2^{-1}. The resulting analytic part of the phase is

$$g_2(p) = i(pr + \xi_1 z') \ , \tag{4.7.43}$$

and its expansion is

$$g_2(p) \simeq g_{22} + g_{22'}(p - c_2^{-1}) \ , \tag{4.7.44}$$

with $g_{22} = g_2(c_2^{-1})$ and $g_{22}' = g_2'(c_2^{-1})$.

For $U(2)$ the amplitude of the integrand in (4.7.21) is analytic and the result can be written as

$$U(2) \approx F_2 \exp(\omega g_{22}) \int_{C_2} \exp\Big\{ \omega \Big[g_{22}'(p - c_2^{-1}) + (2/c_2)^{1/2}|z|(c_2^{-1} - p)^{1/2} \Big] \Big\} dp \ , \tag{4.7.45}$$

where

$$F_2 = \Big\{ \xi_1 E(p) \Big\}_{p = c_2^{-1}} \ . \tag{4.7.46}$$

The branch cut integral can again be viewed as a vertical contour integral above the branch point c_2^{-1}. Introducing a transformation analogous to (4.7.26)

$$c_2^{-1} - p = -ix \ , \tag{4.7.47}$$

maps it into a loop integral around the real axis. Evaluation of this integral yields

$$U(2) \approx iF_2 \exp(\omega g_{22}) \int_0^\infty e^{-a_2 x} \left(e^{-b_2 x^{1/2}} - e^{b_2 x^{1/2}} \right) dx \ , \tag{4.7.48}$$

where

$$a_2 = -i\omega g_{22}' \ , \tag{4.7.49}$$

Table 4.2. Representative quadrant points and their angle definitions on the second sheet. The angle α is increased by 2π and the sheet is defined by continuation through the branch cut starting at $1/c$. Using (4.3.23) this illustrates that $\xi_i < 0$ on this unphysical sheet.

	POINTS	α	β	$(\alpha+\beta)/2$	$\cos\left[(\alpha+\beta/2\right]$	$\sin\left[(\alpha+\beta)/2\right]$
1.	$p_r = 1/c$ $p_i = 2/c$	$7\pi/2$	$\pi/4$	$15\pi/8$	>0	<0
2.	$p_r = -1/c$ $p_i = 2/c$	$7\pi/4$	$\pi/2$	$9\pi/8$	<0	<0
3.	$p_r = -1/c$ $p_i = -2/c$	$9\pi/4$	$3\pi/2$	$15\pi/8$	>0	<0
4.	$p_r = 1/c$ $p_i = -2/c$	$5\pi/2$	$-\pi/4$	$9\pi/8$	<0	<0

and

$$b_2 = (2i/c_2)^{1/2}\omega|z| \ . \tag{4.7.50}$$

The further transformation

$$a_2 x = u^2 \ , \tag{4.7.51}$$

and the expansion of the integrand for large r yields

$$U(2) \sim U_2\omega^{-1/2}r^{-3/2}\exp\left[i\omega(r/c_2 + z'K/c_2)\right] \ , \tag{4.7.52}$$

where

$$U_2 = -2i(2\pi i)^{1/2}\rho_1|z|\rho_2^{-2} \ . \tag{4.7.53}$$

This has the same algebraic decay as $V(1)$ (where the amplitude of the integrand was also analytic) but a different (real) wave shape since the analytic part of the phase is being evaluated at the lower branch point c_2^{-1}. All terms in the evaluation are real.

The evaluation of $V(2)$ is analogous to that of $U(1)$ since its integrand is not analytic at the branch point. The analysis for the evaluation is similar and the result is

$$V(2) \sim V_2 \omega^{-3/2} r^{-3/2} \exp\Big[i\omega(r/c_2 + z'K/c_2)\Big] \ , \tag{4.7.54}$$

where

$$V_2 = -(2\pi i)^{1/2} 2\rho_1 c_2/\rho_2^2 K^2 \ . \tag{4.7.55}$$

This is the same wave shape as (4.7.52) but with a faster algebraic decay in frequency. The result for (4.7.20) is

$$BLI(2) \sim B_2 \omega^{-1/2} r^{-3/2} \exp\Big[i\omega(r/c_2 + z'K/c_2)\Big] \ , \tag{4.7.56}$$

where

$$B_2 = \rho_2 U_2 - \rho_1 V_2 \,\omega^{-1} \ . \tag{4.7.57}$$

Finally combining (4.7.41) and (4.7.56) we get for the transmitted field from (4.7.18) that for $c_2 > c_1$ (using order symbols)

$$\psi_t \sim \exp\Big[i\omega r/c_1 - \omega|z|K/c_2\Big] O(r^{-2})$$

$$+ \exp\Big[i\omega r/c_2 + i\omega z'K/c_2\Big] O(r^{-2}) \ , \tag{4.7.58}$$

consisting of algebraically decaying waves, one of which also decays exponentially in receiver depth. The latter is a fast wave (speed c_1) in range. The second non-decaying wave is slow (speed c_2) and its travel time is $r/c_2 + z'K/c_2$.

4. Appendix

4.A Steepest Descent Method

We describe here a procedure to approximately evaluate integrals of the form

$$I(\omega) = \int_C F(p) e^{\omega f(p)} dp \ , \tag{4.A.1}$$

for large ω. We assume ω is real and positive. The contour C is a contour in the complex p-plane such that the integrand vanishes at the ends of the contour. The problem is that for large values of ω the term $\exp[i\omega \mathrm{Im} f(p)]$ fluctuates rapidly. This makes the evaluation difficult. What saves us is that $F(p)$ varies slowly. We schematically illustrate this in Fig. 4.A.1.

The contour C will go through regions where $\mathrm{Re}\, f(p) > 0$ and $\mathrm{Re}\, f(p) < 0$. The former regions are the most important since the integrand is largest there. What we want to do is to require that in the regions where $\mathrm{Re}\, f(p) > 0$, we can choose $\mathrm{Im}\, f(p)$ equal to a constant so that there are no oscillations.

The point is this: $\exp[i\omega \mathrm{Im}\, f(p)]$ is rapidly oscillating. Its oscillations are a function of the path we take in the complex p-plane. What we want to do is pick out the large value(s) of the integrand and evaluate the integral using these large values. We must somehow fix the contour so that away from the large values the integrand is small, exponentially decreasing if possible.

We write

$$p = p_r + i p_i \ , \tag{4.A.2}$$

and

$$f(p) = u(p) + iv(p) \ . \tag{4.A.3}$$

The function $u(p)$ defines a surface. A *direction of descent* at any point p_s of this surface is a direction in which u decreases from its value at p_s. The resulting contour describes a *path of descent*. We want this

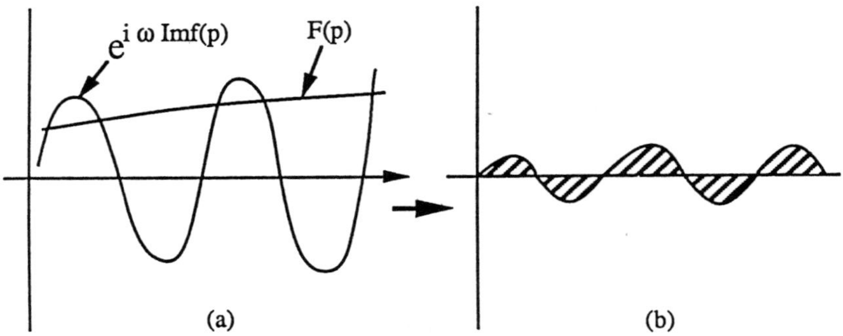

Fig. 4.A.1. In (a) the rapidly oscillating phase and slowly varying amplitude $F(p)$ lead to an integrand in (b) whose contributions cancel.

direction of descent to be the optimal one such that the function falls off as fast as possible. This is called the *path of steepest descent.* The direction of steepest descent (and steepest ascent) about p_s are the directions tangent to $\operatorname{Im} f(p) = v(p) =$ constant. That is $v(p) = v(p_s)$. The proof is that

$$\begin{aligned} \delta f &= f(p) - f(p_s) \ , \\ &= \delta u + i\delta v \ , \end{aligned} \tag{4.A.4}$$

and

$$|\delta f|^2 = |\delta u|^2 + |\delta v|^2 \ . \tag{4.A.5}$$

The change δu is a maximum when $\delta v = 0$, i.e. when v is constant.

We next expand $f(p)$ about p_s in a Taylor series

$$f(p) = f(p_s) + f'(p_s)(p - p_s) + \frac{1}{2} f''(p_s)(p - p_s)^2 + \cdots \ , \tag{4.A.6}$$

and we choose the point p_s by setting

$$f'(p_s) = 0 \ . \tag{4.A.7}$$

With this choice p_s is called the *saddle point.* (It is in fact a saddle point of first order since we also assume that $f''(p_s) \neq 0$.) If f were a real function, (4.A.7) would define a maximum or minimum of f. Since f is complex, u and v satisfy Cauchy-Riemann equations and are harmonic functions, e.g.

$$\frac{\partial^2 u}{\partial p_r^2} + \frac{\partial^2 u}{\partial p_i^2} = 0 \ . \tag{4.A.8}$$

The two curvatures of u thus have opposite sign and at the point p_s, the surface u has the shape of a saddle. We drop higher order terms in (4.A.6) and thus have

$$f(p) \sim f(p_s) + \frac{1}{2} f''(p_s)(p - p_s)^2 \ . \tag{4.A.9}$$

Near $p = p_s$ we can write

$$p - p_s = \rho e^{i\alpha}, \quad \frac{dp}{d\rho} = e^{i\alpha} \ . \tag{4.A.10}$$

Using (4.A.9) and (4.A.10) in (4.A.1) we have that

$$I(\omega) \sim e^{\omega f(p_s)} \int_C F(p) e^{i\alpha} e^{1/2\omega f''(p_s)\rho^2 e^{2i\alpha}} d\rho \ . \tag{4.A.11}$$

In addition, since $F(p)$ is slowly varying it can be approximated by its value at p_s (one term Taylor expansion) and taken outside the integral of (4.A.11) to yield

$$I(\omega) \sim e^{\omega f(p_s)} F(p_s) \int_C e^{i\alpha} e^{1/2\omega f''(p_s)\rho^2 e^{2i\alpha}} d\rho \ . \tag{4.A.12}$$

The most rapid decrease of the integrand is a Gaussian. We have that ($f'' = u'' + iv''$)

$$\begin{aligned} f''(p_s)e^{2i\alpha} &= u''(p_s)\cos 2\alpha - v''(p_s)\sin 2\alpha \\ &\quad + i\left[u''(p_s)\sin 2\alpha + v''(p_s)\cos 2\alpha\right] \end{aligned} \tag{4.A.13}$$

and we can choose the phase oscillations of the integrand of (4.A.12) to vanish if we let the imaginary part of (4.A.13) vanish, i.e. if we choose the angle $\alpha = \alpha_s$ where

$$\tan 2\alpha_s = -v''(p_s)/u''(p_s) \ . \tag{4.A.14}$$

The *angle α_s is the angle the contour makes as it goes through the saddle point.* Using (4.A.14), (4.A.13) can be written as

$$\begin{aligned} f''(p_s)e^{2i\alpha_s} &= -\frac{\sin 2\alpha_s}{v''(p_s)} \left\{ (u''(p_s))^2 + (v''(p_s))^2 \right\} \\ &= -\frac{\sin 2\alpha_s}{v''(p_s)} |f''(p_s)|^2 \ . \end{aligned} \tag{4.A.15}$$

Equation (4.A.12) thus becomes

$$I(\omega) \sim e^{\omega f(\rho_s)} F(\rho_s) e^{i\alpha_s} \int_C e^{-a\rho^2} d\rho \ , \qquad (4.A.16)$$

where

$$a = \frac{\omega \sin 2\alpha_s}{2v''(p_s)} |f''(p_s)|^2 \ . \qquad (4.A.17)$$

If $\operatorname{sgn} v''(p_s) = +1$, the maximum decay of the Gaussian in (4.A.16) occurs if $\sin 2\alpha_s = 1$ or $\alpha_s = \pi/4$. If $\operatorname{sgn} v''(p_s) = -1$, the maximum decay occurs if $\sin 2\alpha_s = -1$ or $\alpha_s = -\pi/4$. Both cases can be treated if we write (4.A.16)as

$$I(\omega) \sim e^{\omega f(p_s)} F(p_s) e^{i sgn[v''(p_s)]\frac{\pi}{4}} \int_C e^{-a_M \rho^2} d\rho \ , \qquad (4.A.18)$$

where

$$a_M = \frac{\omega}{2} \frac{|f''(p_s)|^2}{|v''(p_s)|} \ . \qquad (4.A.19)$$

Finally, to evaluate the integral in (4.A.18), we assume that it can be treated as the integral of a Gaussian where the path is extended from $-\infty$ to ∞, since the extension adds a negligible amount to the integral, that is

$$\int_C e^{-a_M \rho^2} d\rho \simeq \int_{-\infty}^{\infty} e^{-a_M \rho^2} d\rho = \left(\frac{\pi}{a_M}\right)^{1/2} \ . \qquad (4.A.20)$$

The final result from (4.A.18) becomes

$$I(\omega) \sim e^{\omega f(p_s)} F(p_s) e^{i \operatorname{sgn}[v''(p_s)]\frac{\pi}{4}} \left\{ \frac{2\pi |v''(p_s)|}{\omega |f''(p_s)|^2} \right\}^{1/2} \ . \qquad (4.A.21)$$

As an example, consider the phase function for the reflected field in this chapter,

$$f(p) = i\left[pr + \xi_1(z + z')\right] ; \quad \xi_1 = \left(c_1^{-2} - p^2\right)^{1/2} \ . \qquad (4.A.22)$$

The saddle point occurs when

$$f'(p_s) = 0 \ , \qquad (4.A.23)$$

or when

$$r\xi_{1s} = p_s(z + z') \ . \qquad (4.A.24)$$

This can be parametrized as

$$p_s = c_1^{-1} \sin \theta_s \, ; \quad \xi_{1s} = c_1^{-1} \cos \theta_s \ , \tag{4.A.25}$$

so that $p_s < c_1^{-1}, \xi_{1s}$ is real, and

$$\tan \theta_s = r/(z+z') \ . \tag{4.A.26}$$

We also have

$$f''(p) = -i(z+z')/\xi_1^3 c_1^2 \ , \tag{4.A.27}$$

so that at the saddle point

$$\begin{aligned} f(p_s) &= ip_s[r^2+(z+z')^2]/r \\ &= iR^2 \sin\theta_s/c_1 r \ , \quad \left(R = [r^2+(z+z')^2]^{1/2}\right) \ , \\ &= iR/c_1 \ , \end{aligned} \tag{4.A.28}$$

$$\begin{aligned} |f''(p_s)| &= |v''(p_s)| \\ &= (z+z')/\xi_{1s} cos^2\theta_s \\ &= R^2/\xi_{1s}(z+z') \ , \end{aligned} \tag{4.A.29}$$

and $\mathrm{sgn} v''(p_s) = -1$ so that $\alpha_s = -\pi/4$.

From (4.A.21) we thus have

$$I(\omega) \sim F(p_s) e^{-i\pi/4} e^{i\omega R/c_1} R^{-1} \left\{ 2\pi \xi_{1s}(z+z')/\omega \right\}^{1/2} \ ,$$

which appears as a modified spherical wave.

5. Two-Layered Liquid Half-Space (Pekeris Waveguide)

We consider an example of a waveguide propagation problem which is not only formally solvable but whose solution contains a rich vein of formal and approximate analysis interspersed with physical interpretations. It is truly a canonical problem.

5.1 Geometry and Notation

We consider the propagation of sound in a two-dimensional (range and depth) two-layer liquid half-space usually referred to as the Pekeris waveguide [5.1]. The geometry is illustrated in Fig. 5.1.

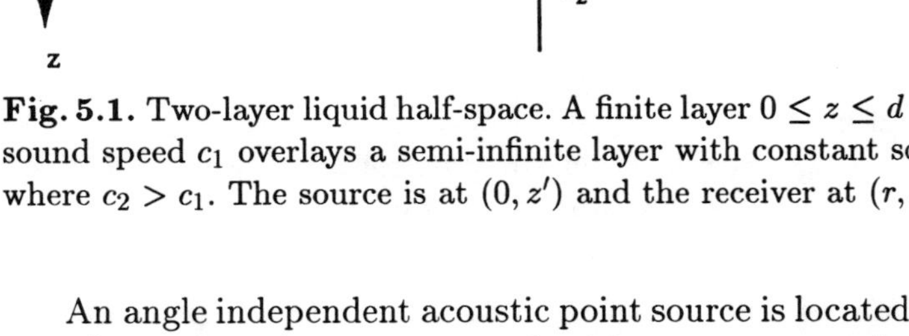

Fig. 5.1. Two-layer liquid half-space. A finite layer $0 \le z \le d$ with constant sound speed c_1 overlays a semi-infinite layer with constant sound speed c_2 where $c_2 > c_1$. The source is at $(0, z')$ and the receiver at (r, z).

An angle independent acoustic point source is located at $r = 0$ and $z = z'$ in a finite layer, $0 \le z \le d$, with constant sound speed c_1. The receiver is located in the same layer at (r, z). The finite layer overlays a semi-infinite half space, $z > d$, with constant sound speed c_2 where $c_2 > c_1$. For this simple model no density variations are considered although these could be easily included. The boundary value problem is analogous to that considered in Chap. 4 for a point source over a half-space (here with no density variation) but now with the important added requirement of an upper boundary at $z = 0$. The point source solution was an integral, i.e. a continuous spectrum of waves. The existence of an upper boundary here admits the possibility of an additional discrete spectrum, i.e. waveguide modes. For perfectly reflecting boundaries at $z = 0$ and $z = d$, only the discrete spectrum would exist.

The boundary value problem we describe can be considered as a simple example describing sound propagation in the ocean waveguide overlaying a semi-infinite bottom. It is "simple" for several reasons, the main one being the fact that in the real ocean c_1 is a function of depth (due to temperature, pressure and salinity variability of the medium) with a minimum value in the waveguide. The problem does, however, correspond well to propagation in shallow water and illustrates features which appear in the deep ocean problem.

5.2 Green's Function

We have a two-dimensional cylindrically symmetric problem with cylindrical coordinates r and z. The Green's function $G(\boldsymbol{r}, \boldsymbol{r}')$ satisfies the two-dimensional Helmholtz equation

$$\frac{1}{r}\frac{\partial}{\partial r}\left(r\frac{\partial G}{\partial r}\right) + \frac{\partial^2 G}{\partial z^2} + k_0^2 K^2(z) G = -\frac{\delta(r)\delta(z-z')}{2\pi r} , \qquad (5.2.1)$$

for a point source located at $\boldsymbol{r}' = (0, z')$ and a field point $\boldsymbol{r} = (r, z)$. (We simplify notation slightly from Chap. 4 in that here the transverse source position is zero.) Here $k_0 = \omega/c_0$ where c_0 is some reference sound speed (usually taken at the surface) and $K(z) = c_0/c(z)$ where $c(z)$ is the depth-dependent sound speed for the full two-layer problem, and is specified later.

We assert that, since K is only a function of z, we can write a separable solution of (5.2.1) in the spectral form

$$G(\boldsymbol{r}, \boldsymbol{r}') = \frac{k_0^2}{2\pi i}\int_{\Gamma_\lambda} G_1(r;\lambda) G_2(z, z';\lambda) d\lambda , \qquad (5.2.2)$$

where Γ_λ encloses all the singularities of $(G_2)G_1$ on the proper sheet of the complex λ-plane in a (counter-) clockwise direction. Here λ is the separation parameter and G_1 and G_2 satisfy the equations

$$\frac{1}{r}\frac{d}{dr}\left(r\frac{dG_1}{dr}\right) = -k_0^2\lambda G_1 - \frac{\delta(r)}{2\pi r} \ , \tag{5.2.3}$$

and

$$\frac{d^2G_2}{dz^2} + k_0^2K^2(z)G_2 = k_0^2\lambda G_2 - \delta(z-z) \ . \tag{5.2.4}$$

We prove this assertion in Appendix 5A. This is a somewhat more general procedure for finding the separable representation for the Green's function than we have used up to now. The reason for its introduction is that because of the waveguide nature of the problem G_2 will contain additional singularities not present before. We will express our result as the counter-clockwise integral Γ_2 around all these singularities of G_2.

To find G_1 and G_2 we must specify the boundary conditions. We assume that as $r \to \infty$, G_1 must behave as an outgoing cylindrical wave. The solution of (5.2.3) is thus (see Chap. 2) with $0 < \arg\lambda < 2\pi$

$$G_1(r;\lambda) = \left(\frac{i}{4}\right) H_0^{(1)}(k_0\sqrt{\lambda}r) \ . \tag{5.2.5}$$

Further if we define $\lambda = q^2$, then

$$G_1(r;q^2) = \left(\frac{i}{4}\right) H_0^{(1)}(k_0 qr) \ . \tag{5.2.6}$$

For an outgoing (or decaying) wave, we require $\mathrm{Im}q > 0$, so that in the complex q–plane we are restricted to the upper half-plane. We used the same solution in Chap. 4 and, comparing results, we see that q is the dimensionless ray parameter, $q = c_0 p$.

To specify G_2 we note that from (5.2.4) it satisfies

$$\frac{d^2G_2}{dz^2} + k_0^2\left[K^2(z) - q^2\right] G_2 = -\delta(z-z) \ . \tag{5.2.7}$$

We specify the function $K(z)$ as

$$K(z) = \begin{cases} K_1 = c_0/c_1, & 0 \le z \le d \\ \\ K_2 = c_0/c_2, & z > d \end{cases} \tag{5.2.8}$$

where each K_j is constant. In the interior (i) of the waveguide $(0 \le z \le d)$ we assume a pressure release (soft, Dirichlet) boundary condition at $z = 0$,

$$G_2^i(0,z';q^2) = 0 \ , \tag{5.2.9}$$

so that the general solution can be written as

$$G_2^i(z, z'; q^2) = A_i \sin(k_0\mu_1 z_<) \cdot \left\{\sin(k_0\mu_1 z_<) + B_i \cos(k_0\mu_1 z_>)\right\} \ , \qquad (5.2.10)$$

where A_i and B_i are constants to be specified by the boundary conditions, $z_<$ and $z_>$ are the lesser and greater of z and z', and

$$\mu_1 = (K_1^2 - q^2)^{1/2} \ . \qquad (5.2.11)$$

With the choice (5.2.10) it is easily seen that we have already satisfied the continuity of G_2^i at $z = z'$. The jump condition

$$\lim_{\varepsilon\to 0+} \left\{ \left.\frac{dG_2^i}{dz}\right|_{z=z'+\varepsilon} - \left.\frac{dG_2^i}{dz}\right|_{z=z'-\varepsilon} \right\} = -1 \ , \qquad (5.2.12)$$

is used to relate A_i and B_i. Substituting (5.2.10) into (5.2.12) we get

$$A_i B_i k_0 \mu_1 = 1 \ . \qquad (5.2.13)$$

In the exterior (e) of the waveguide ($z > d$), the solution of (5.2.7) with (5.2.8) is

$$G_2^e(z, z; q^2) = A_e e^{ik_0\mu_2(z-d)} \ , \qquad (5.2.14)$$

where

$$\mu_2 = (K_2^2 - q^2)^{1/2} \ , \qquad (5.2.15)$$

and we require

$$\mathrm{Re}\,\mu_2 > 0, \quad \mathrm{Im}\mu_2 > 0 \ , \qquad (5.2.16)$$

in order to insure that G_2^e satisfies an outgoing radiation condition as $z \to \infty$. The constant A_e must be determined.

We require G_2^i and G_2^e to satisfy the following continuity conditions at $z = d$

$$G_2^i(d, z'; q^2) = G_2^e(d, z'; q^2) \ , \qquad (5.2.17)$$

and

$$\frac{d}{dz} G_2^i(d, z'; q^2) = \frac{d}{dz} G_2^e(d, z; q^2) \ . \qquad (5.2.18)$$

These follow from the fact that G is the Green's function corresponding to a velocity potential, so the physical continuity conditions are thus those of pressure (which is (5.2.17) here with equal densities) and the normal component of velocity which is (5.2.18). Substituting (5.2.10) and (5.2.14) in (5.2.17) and (5.2.18) and using (5.2.13) we find

$$B_i = D(q)/N(q) \ , \tag{5.2.19}$$

$$A_i = N(q)/k_0\mu_1 D(q) \ , \tag{5.2.20}$$

and

$$A_e = \sin(k_0\mu_1 z')/k_0 D(q) \ , \tag{5.2.21}$$

where

$$N(q) = \mu_1 \sin(k_0\mu_1 d) + i\mu_2 \cos(k_0\mu_1 d) \ , \tag{5.2.22}$$

and

$$D(q) = \mu_1 \cos(k_0\mu_1 d) - i\mu_2 \sin(k_0\mu_1 d) \ . \tag{5.2.23}$$

Note that we were able to unambiguously solve for the remaining constants A_i, B_i and A_e without any additional source amplitude. This was not because we had an extra boundary condition but rather because we specified the source strength as unity in (5.2.1).

Substituting (5.2.19) and (5.2.20) in (5.2.10) we can write the result as

$$G_2^i(z, z'; q^2) = \frac{\sin(k_0\mu_1 z_<) N_1(q)}{k_0\mu_1 D(q)} \ , \tag{5.2.24}$$

where

$$N_1(q) = \mu_1 \cos\left[k_0\mu_1(d - z_>)\right] - i\mu_2 \sin\left[k_0\mu_1(d - z_>)\right] \ . \tag{5.2.25}$$

Similarly, the full exterior solution is from (5.2.14) and (5.2.21)

$$G_2^e(z, z'; q^2) = \frac{\sin(k_0\mu_1 z') e^{ik_0\mu_2(z-d)}}{k_0 D(q)} \ . \tag{5.2.26}$$

Finally, using (5.2.6) in (5.2.2) we get that ($\lambda = q^2$)

$$G(\boldsymbol{r}, \boldsymbol{r}') = \frac{k_0^2}{4\pi} \int_C H_0^{(1)}(k_0 q r) G_2(z, z; q^2) q dq \ , \tag{5.2.27}$$

where the contour C encloses all the singularities of G_2 in a counter-clockwise direction and on the proper Riemann sheet of the complex q–plane. We must determine these singularities next. The contour C can be easily seen to follow from the representation (5.2.2) where we choose the contour Γ_1 enclosing the singularities of G_1 or $H_0^{(1)}(k_0\sqrt{\lambda} r)$ in a clockwise direction. The only singularity is a branch point at $\lambda = 0$, and Γ_1 is illustrated in Fig. 5.2. Using $\lambda = q^2$, the contour C follows and is also illustrated in Fig. 5.2.

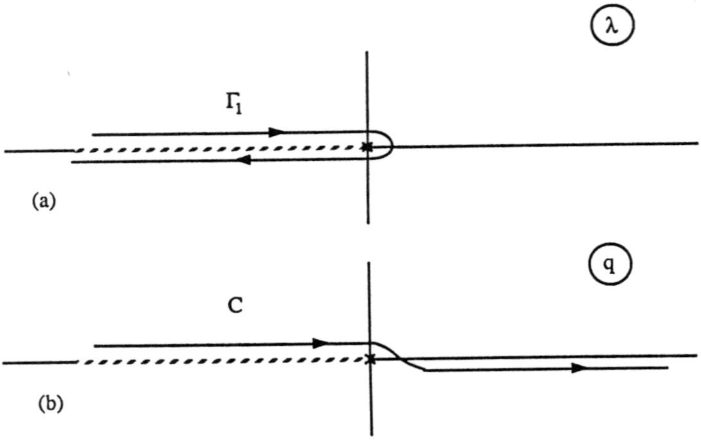

Fig. 5.2. (a) Contour Γ_1 enclosing the singularities of G_1 in a clockwise direction in the λ-plane. G_1 is the Hankel function with branch points at 0 and ∞. (b) Contour C in the complex q-plane with $\lambda = q^2$. The contour encloses the singularities of G_2 with Im$q > 0$ to satisfy the radiation condition.

5.3 Analytic Properties of G_2

In order to evaluate the integral representation (5.2.27) we need to know the analytic properties of the integrand as a function of q. In particular the contour C will enclose those singularities of G_2 in the upper half of the complex q-plane. This region also insures that the radial radiation condition is satisfied since $q_i > 0$. From the representations (5.2.24) and (5.2.26) it immediately appears that G_2 has possible singularities at: (a) branch points at $q = \pm K_1$ from the vanishing of μ_1; (b) branch points at $q = \pm K_2$ from the vanishing of μ_2 and (c) poles which follow from the zeroes of D. We must now investigate whether in fact these are indeed singularities of G_2. We define the square root branch cuts in exactly the same manner as in Chap. 4. Their orientation is as discussed in Sect. 4.3. For a generic value K, we have

$$K - q = |K - q| \exp(i\alpha) \doteq R_+ \exp(i\alpha) \ , \tag{5.3.1}$$

and

$$K + q = |K + q| \exp(i\beta) \doteq R_- \exp(i\beta) \ , \tag{5.3.2}$$

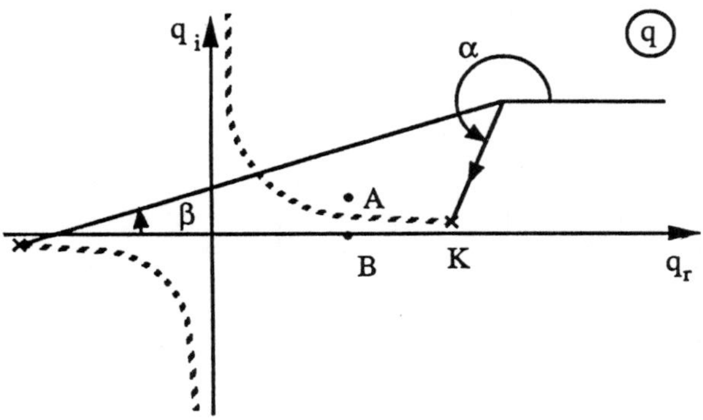

Fig. 5.3. Square root branch cuts and angle definitions in the complex q-plane, with $q = c_0 p$ where p is the ray parameter. The orientation of the cuts is defined in Chap. 4 where many angle values are listed. At point A, $\alpha + \beta = 2\pi$. At point B, $\alpha + \beta = 0$.

so that

$$\mu = (K^2 - q^2)^{1/2} = (R_+ R_-)^{1/2} \exp\left[i(\alpha + \beta)/2\right] . \tag{5.3.3}$$

The angles α and β are defined in Fig. 5.3. Recall from our discussion in Chap. 4 that on the first sheet of this Riemann surface, the imaginary part of the square root is positive, i.e. $\mu_i > 0$. (In Chap. 4 the square root ξ is just μ / c_0 here).

(a) Branch Cuts at $q = \pm K_1$ For specific branch cuts at $q = \pm K_1$ define the discontinuity of G_2 across these branch cuts as

$$\Delta_1 G_2(z, z'; q^2) = G_2(z, z'; q^2)\Big|_{\alpha_1 + \beta_1 = 0} - G_2(z, z'; q^2)\Big|_{\alpha_1 + \beta_1 = 2\pi} , \tag{5.3.4}$$

so that μ_2 is analytic and

$$\mu_1 = \begin{cases} |\mu_1|, & \alpha_1 + \beta_1 = 0, \\ -|\mu_1|, & \alpha_1 + \beta_1 = 2\pi \end{cases} . \tag{5.3.5}$$

Using (5.2.24) we see that

$$G_2^i(z, z'; q^2)\Big|_{\alpha_1 + \beta_1 = 0} = \frac{\sin(k_0 |\mu_1| z_<)}{k_0 |\mu_1|} \left[\frac{N_1(q)}{D(q)}\right]_{\alpha_1 + \beta_1 = 0} , \tag{5.3.6}$$

where the numerator (N_1) and denominator (D) terms are

$$\left[\frac{N_1(q)}{D(q)}\right]_{\alpha_1+\beta_1=0} = \frac{|\mu_1|\cos\left[k_0|\mu_1|(d-z_>)\right] - i\mu_2\sin\left[k_0|\mu_1|(d-z_>)\right]}{|\mu_1|\cos(k_0\mu_1 d) - i\mu_2\sin(k_0|\mu_1|d)} \quad . \tag{5.3.7}$$

(Note that the absolute value is unimportant inside the argument of a cosine which is an even function. The sine is odd however and it does matter here). Further we see that

$$G_2^i(z,z';q^2)\Big|_{\alpha_1+\beta_1=2\pi} = \frac{\sin(k_0|\mu_1|z_<)}{k_0|\mu_1|}\left[\frac{N_1(q)}{D(q)}\right]_{\alpha_1+\beta_1=2\pi} \quad , \tag{5.3.8}$$

where

$$\left[\frac{N_1(q)}{D(q)}\right]_{\alpha_1+\beta_1=2\pi} = \frac{-|\mu_1|\cos\left[k_0|\mu_1|(d-z_>)\right] + i\mu_2\sin\left[k_0|\mu_1|(d-z_>)\right]}{-|\mu_1|\cos(k_0\mu_1 d) + i\mu_2\sin(k_0|\mu_1|d)} \quad . \tag{5.3.9}$$

But (5.3.9) is equal to (5.3.7), so at least for G_2^i there is no discontinuity. Using (5.2.26) we can easily show that the same result holds for G_2^e. Thus from (5.3.4)

$$\Delta_1 G_2(z,z';q^2) = 0 \quad , \tag{5.3.10}$$

and in fact G_2 does *not* have a branch point at $q = +K_1$. The same result is of course true at $q = -K_1$. This is already a different result from that in Chap. 4 where the vanishing square root did lead to branch points. The result here can also be easily seen to follow by noting that G_2 is an even function of μ_1.

(b) Branch Cuts at $q = \pm K_2$ For branch cuts at $q = \pm K_2$ define the discontinuity of G_2 across these branch cuts as

$$\Delta_2 G_2(z,z';q^2) = G_2(z,z';q^2)\Big|_{\alpha_2+\beta_2=0} - G_2(z,z';q^2)\Big|_{\alpha_2+\beta_2=2\pi} \quad , \tag{5.3.11}$$

so that μ_1 is analytic and

$$\mu_2 = \begin{cases} |\mu_2|\,, & \alpha_2+\beta_2=0\,, \\ -|\mu_2|\,, & \alpha_2+\beta_2=2\pi\,. \end{cases} \tag{5.3.12}$$

Using (5.2.24) we can see that (5.3.11) follows if we know

$$\Delta_2\left[\frac{N_1(q)}{D(q)}\right] = \left.\frac{N_1(q)}{D(q)}\right|_{\alpha_2+\beta_2=0} - \left.\frac{N_1(q)}{D(q)}\right|_{\alpha_2+\beta_2=2\pi} \,. \tag{5.3.13}$$

From (5.2.25) and (5.2.23) these latter two terms are complex conjugates so that we get

$$\Delta_2\left[\frac{N_1(q)}{D(q)}\right] = 2i\mathrm{Im}\left\{\left.\frac{N_1(q)}{D(q)}\right|_{\alpha_2+\beta_2=0}\right\} \,, \tag{5.3.14}$$

where Im represents the imaginary part. The explicit result is

$$\Delta_2\left[\frac{N_1(q)}{D(q)}\right] = \frac{N_2(q)}{D_2(q)} \,, \tag{5.3.15}$$

where

$$N_2(q) = 2i\mu_1|\mu_2|\sin(k_0\mu_1 z_>) \,, \tag{5.3.16}$$

and

$$\begin{aligned} D_2(q) &= \mu_1^2\cos^2(k_0\mu_1 d) + |\mu_2|^2\sin^2(k_0\mu_1 d) \,, \\ &= D(q)D^*(q) \,. \end{aligned} \tag{5.3.17}$$

The full discontinuity of the interior solution is thus

$$\Delta_2 G_2^i(z,z';q^2) = \frac{2i|\mu_2|\sin(k_0\mu_1 z_<)\sin(k_0\mu_1 z_>)}{k_0 D_2(q)} \,. \tag{5.3.18}$$

From (5.2.26) we can write the discontinuity of the exterior solution as

$$\Delta_2 G_2^e(z,z';q^2) = \frac{\sin(k_0\mu_1 z')}{k_0}\Delta_2\left[\frac{e^{ik_0\mu_2(z-d)}}{D(q)}\right] \,. \tag{5.3.19}$$

The latter discontinuity can also be written as

$$\Delta_2\left[\frac{e^{ik_0\mu_2(z-d)}}{D(q)}\right]=2i\,\mathrm{Im}\left[\frac{e^{ik_0\mu_2(z-d)}}{D(q)}\right]_{\alpha_2+\beta_2=0}\quad, \tag{5.3.20}$$

and is explicitly given by

$$\Delta_2\left[\frac{e^{ik_0\mu_2(z-d)}}{D(q)}\right]=\frac{N_3(q)}{D_2(q)}\quad, \tag{5.3.21}$$

where

$$\begin{aligned} N_3(q) &= 2i\mu_1\sin\left[k_0|\mu_2|(z-d)\right]\cos(k_0\mu_1 d) \\ &\quad +2i|\mu_2|\sin(k_0\mu_1 d)\cos\left[k_0|\mu_2|(z-d)\right]\quad, \end{aligned} \tag{5.3.22}$$

and D_2 is given by (5.3.17). The final result is, from (5.3.19),

$$\Delta_2 G_2^e(z,z';q^2)=\frac{N_3(q)\sin(k_0\mu_1 z')}{k_0 D_2(q)}\quad, \tag{5.3.23}$$

which clearly differs from the discontinuity of the interior solution (5.3.18).

(c) Poles on the First Sheet The poles of G_2 arise from the zeroes of the denominator term $D(q)$. Here we consider only those possible poles on the first sheet, since these are the only poles we surround with our contour representation (5.2.27). (We discuss this further in later sections). Set (5.2.23) to zero and write the result as ($\mu_1 \neq 0$)

$$\cot\left(k_0\mu_1 d\right)=i\mu_2/\mu_1\quad, \tag{5.3.24}$$

where following (5.3.3) we can write

$$\mu_1=|\mu_1|\left\{\cos\left[\frac{1}{2}(\alpha_1+\beta_1)\right]+i\sin\left[\frac{1}{2}(\alpha_1+\beta_1)\right]\right\}\quad, \tag{5.3.25}$$

and

$$\mu_2=|\mu_2|\left\{\cos\left[\frac{1}{2}(\alpha_2+\beta_2)\right]+i\sin\left[\frac{1}{2}(\alpha_2+\beta_2)\right]\right\}\quad. \tag{5.3.26}$$

Note that even though G_2 has no branch point when μ_1 vanishes (as we showed in part (a)), the denominator function $D(q)$ does have branch

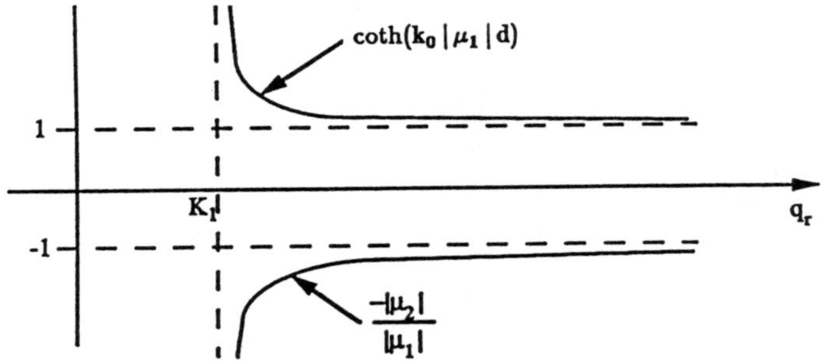

Fig. 5.4. Sketch of the two sides of (5.3.27) as a function of q_r illustrating that they do not intersect.

points when $\mu_1 = 0$ as well as when $\mu_2 = 0$, so the square roots must be defined in both cases. We first search for solutions of (5.3.24) by considering various special cases.

Since $D(q)$ and consequently (5.3.24) are functions of q^2 through μ_1 and μ_2 we need only search for solutions of (5.3.24) in the upper half of the q-plane.

1. $q = iq_i, q_i > 0$

   There are *no solutions* of (5.3.24) when q is pure imaginary since on the imaginary axis $\alpha_1+\beta_1 = 0$ and $\alpha_2+\beta_2 = 0$. (Note that the imaginary axis is to the *left* of the branch cuts.) Consequently the *lhs* of (5.3.24) is pure real and the *rhs* pure imaginary.

2. $q = q_r$ and $q_r^2 < K_2^2$

   On the real axis between $\pm K_2$ we again have $\alpha_1 + \beta_1 = 0$ and $\alpha_2 + \beta_2 = 0$. The conclusion is the same as (1) above and there are *no solutions*.

3. $q = q_r$ and $q_r^2 > K_1^2$

   On the real axis for $q_r > K_1$ we have $\alpha_1 = \alpha_2 = \pi$ and $\beta_1 = \beta_2 = 0$ so that from (5.3.25) and (5.3.26) $\mu_1 = i|\mu_1|$ and $\mu_2 = i|\mu_2|$. We can thus write (5.3.24) as

$$\coth\left(k_0|\mu_1|d\right) = -|\mu_2|/|\mu_1| \tag{5.3.27}$$

   which has *no solutions* since the two sides of (5.3.27) fail to intersect. They are sketched in Fig. 5.4.

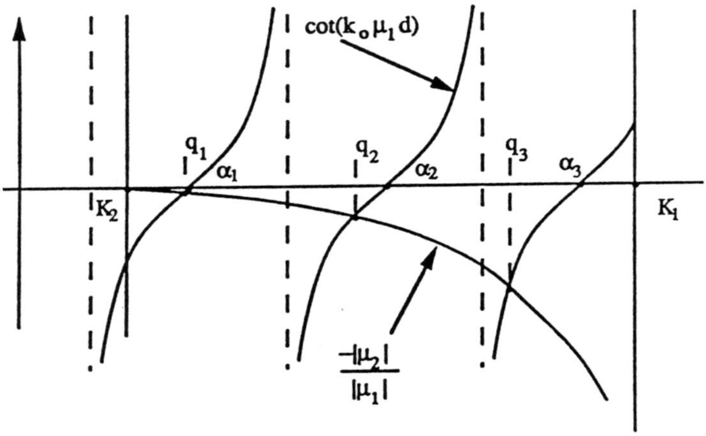

Fig. 5.5. Sketch of the two sides of (5.3.28) showing three intersections occurring at points $q_j (j = 1, 2, 3)$. The zeroes of the cotangent occur at $\alpha_j (j = 1, 2, 3)$.

4. $q = q_r$ and $K_2^2 < q_r^2 < K_1^2$

   On the real axis for $K_2 < q_r < K_1$ we have $\alpha_2 = \pi$, $\beta_2 = 0$ and $\alpha_1 = 0$, $\beta_1 = 0$ so that μ_1 is pure real and $\mu_2 = i|\mu_2|$. Then (5.3.24) becomes

$$\cot(k_0\mu_1 d) = -|\mu_2|/\mu_1 \ , \tag{5.3.28}$$

   which can have solutions $q = q_j (j = 1, 2, \cdots, n)$ provided K_2 and K_1 are appropriately spaced with respect to the period of the cotangent. An example when $n = 3$ is illustrated in Fig. 5.5. Both sides of (5.3.28) are plotted and their points of intersection yield the $\{q_j\}$. *The set* $\{q_j\}$ *constitute the discrete eigenvalue spectrum of our problem and are termed the discrete normal modes or proper modes* of the problem.

   The actual positions q_j must be found numerically. They depend on the values of the sound speeds c_1 and c_2 and the scaled depth $k_0 d$. The zeroes of the cotangent, α_j can be found from

$$\cos\left[k_0(K_1^2 - \alpha_j^2)^{1/2} d\right] = 0 \ , \tag{5.3.29}$$

   or at values α_j given by

$$\alpha_j = \left\{K_1^2 - [(2j+1)\pi/2k_0 d]^2\right\}^{1/2} , \qquad (5.3.30)$$

which must be real. The larger $k_0 d$ and/or K_1 are, the more values of j exist for which α_j is real, $j = 1, 2, \cdots, n$, where the count of these j values also serves to count the number of normal modes.[1] Since D is a function of q^2, poles also occur at values $-K_1 < q_r < -K_2$. These are excluded from our contour integral since negative q_r values would yield Hankel functions which would behave as incoming radial waves and would violate the radiation condition. This is discussed further below. There is one last case to be discussed, however, before we can give an explicit representation of the solution.

Recall that in Chap. 4 we had a semi-infinite medium above the interface, so that the solutions were limited to upgoing waves. Here in the waveguide, the proper mode solutions can be thought of as a combination of both up- and down-going plane waves which interfere in the waveguide and at the interface at $z = d$ and the surface at $z = 0$. *The angle of incidence of these proper modes on either surface is greater than the critical angle of incidence on the interface.* This is because $q_j > K_2$ which implies that $p_j > c_2^{-1}$. The ray parameter is $p_j = c_1^{-1} \sin\theta_j$ where θ_j is the incidence angle for the jth mode. From the inequality we thus get $\sin\theta_j > \sin\theta_c$ so that $\theta_j > \theta_c$. The reason this is true for both surfaces is that the rays bounce back and forth between $z = 0$ and $z = d$ without any further refraction taking place. Hence, the angle of incidence is the same for both surfaces.

5. q complex

We prove below that there are no other complex poles on the first sheet. We discuss the case of other sheets later. We rewrite (5.3.24) as

$$\mu_1 \cos(k_0 d \mu_1) = i\mu_2 \sin(k_0 d \mu_1) , \qquad (5.3.31)$$

and substitute for μ_1 and μ_2 their values in terms of real and imaginary parts

$$\mu_1 = \mu_{1r} + i\mu_{1i} , \qquad \mu_2 = \mu_{2r} + i\mu_{2i} . \qquad (5.3.32)$$

---

[1] Note that α_j is independent of K_2.

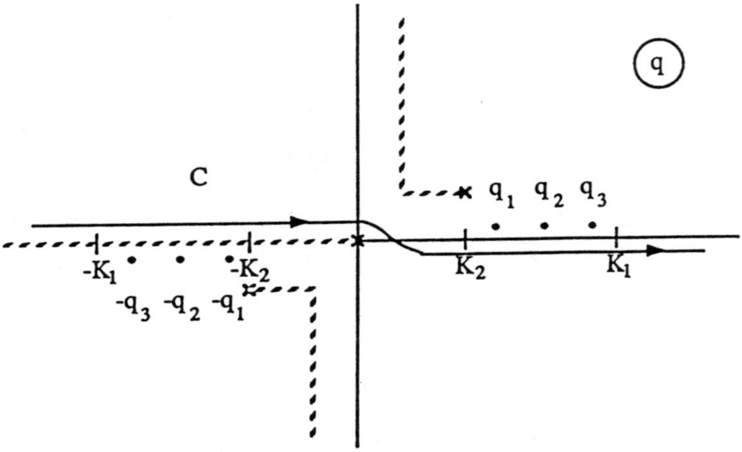

Fig. 5.6. Contour C and the analytic properties of the integrand of (5.4.1) in the complex q-plane. Branch points at $q = 0$ (from the Hankel function) and $q = \pm K_2$ (from G_2) are indicated as well as three proper modes (q_j) arising from G_2. All are slightly offset from the axes for clarity.

On the first sheet both μ_{1i} and μ_{2i} are positive, and as we discussed in Sect. 4.3 both μ_{1i} and μ_{2i} have the same sign in the different regions of the problem, so that $\mu_{1r}\mu_{2r}$ is always positive. Substituting (5.3.32) into (5.3.31) and equating the real and imaginary parts we can write the equations as

$$\mu_{1r} + \mu_{2r} \tanh(k_0 d\mu_{1i}) = -\{\mu_{1i} \tanh(k_0 d\mu_{1i}) + \mu_{2i}\} \tan(k_0 d\mu_{1r}) \ , \qquad (5.3.33)$$

and

$$\mu_{1i} + \mu_{2i} \tanh(k_0 d\mu_{1i}) = \{\mu_{1r} \tanh(k_0 d\mu_{1i}) + \mu_{2r}\} \tan(k_0 d\mu_{1r}) \ . \qquad (5.3.34)$$

Divide the equations and multiply out to get

$$(\mu_{1r}^2 + \mu_{2r}^2 + \mu_{1i}^2 + \mu_{2i}^2) tanh(k_0 d\mu_{1i}) = = -\left[1 + \tanh^2(k_0 d\mu_{1i})\right] [\mu_{1r}\mu_{2r} + \mu_{1i}\mu_{2i}] \ . \qquad (5.3.35)$$

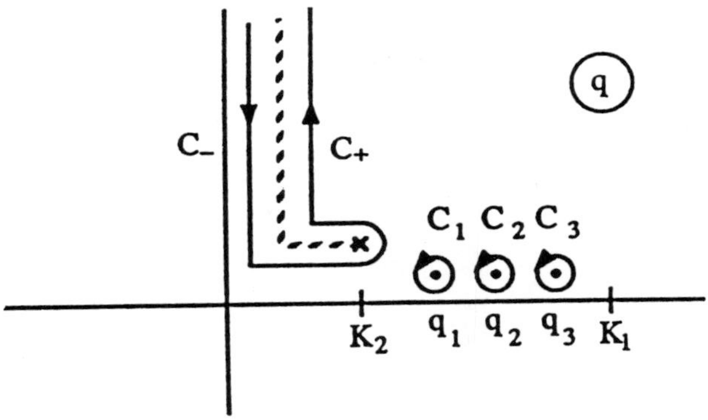

Fig. 5.7. Equivalent contour to that of Fig. 5.6. Illustrated are integrals along the branch line plus closed contours C_j around the poles.

The *lhs* of (5.3.35) is always positive while the *rhs* is always negative. We thus have no other first sheet poles. The normal modes are the only ones. By removing the possibility of other poles we can now distort the contour and write an explicit representation of the solution. (We also note that our normal mode poles which occur when $\mu_{1i} = 0$ and $\mu_{2r} = 0$ satisfy (5.3.35)).

5.4 Normal Mode (NM) Representation

Recall from (5.2.27) that the integral representation for G is

$$G(\boldsymbol{r}, \boldsymbol{r}') = \frac{k_0^2}{4\pi} \int_C H_0^{(1)}(k_0 qr) G_2(z, z'; q^2) q dq \ , \tag{5.4.1}$$

where the contour C encloses all the singularities of G_2 in the upper half of the first sheet of the complex q-plane in a counter-clockwise manner. G_2 has a branch cut at $q = K_2$ and poles at $q_j (j = 1, 2, \cdots, n)$ which are the proper discrete modes. The analytic properties are indicated in Fig.5.6, and the equivalent contour in Fig. 5.7.

Using the contour representations illustrated in Figs. (5.6) and (5.7) we can write (5.4.1) as the sum of two terms, a branch line integral (BLI) contribution plus a residue series (RS) sum over the proper modes. That is,

$$G(\boldsymbol{r},\boldsymbol{r}') = BLI + RS \ , \tag{5.4.2}$$

where

$$BLI = \frac{k_0^2}{4\pi}\int_C H_0^{(1)}(k_0 qr)\Delta_2 G_2(z,z';q^2)q\,dq \ , \tag{5.4.3}$$

with $\Delta_2 G_2$, the discontinuity of G_2 across the cut, defined by (5.3.11) and explicitly given by (5.3.18) or (5.3.23), and where

$$RS = \frac{ik_0^2}{2}\sum_{j=1}^{n} H_0^{(1)}(k_0 q_j r)q_j RES\left[G_2(z,z';q_j^2)\right] \ , \tag{5.4.4}$$

is in terms of the residues (RES) of G_2 at the proper mode poles q_j. *The representation* (5.4.2) *is referred to as the normal mode (NM) representation.* Note that it includes *both* a discrete and a continuous spectrum for this problem. Sturm-Liouville theory guarantees at most a finite number of discrete poles plus a continuous spectrum [5.2].

To evaluate (5.4.4) we must find the residue of G_2 at the proper mode poles q_j. The region of the proper modes corresponds to the region where μ_1 is real and $\mu_2 = i|\mu_2| = i(q^2 - K_2^2)^{1/2}$. The residue is defined as

$$RES\left[G_2(z,z';q_j^2)\right]$$

$$= \lim_{q\to q_j}(q-q_j)\begin{cases} G_2^i(z,z';q^2) \ , & 0\le z\le d \\ G_2^e(z,z';q^2) \ , & z>d \ , \end{cases} \tag{5.4.5}$$

where G_2^i is defined by (5.2.24) and G_2^e by (5.2.26). Both contain the same denominator $D(q)$ defined by (5.2.23), and to solve (5.4.5) for either case we need

$$\lim_{q\to q_j}\frac{(q-q_j)}{D(q)} \ . \tag{5.4.6}$$

Expand $D(q)$ in a Taylor expansion about q_j

$$D(q) = D(q_j) + (q-q_j)\left[\frac{d}{dq}D(q)\right]_{q_j} + \cdots$$

and keep the first two terms. Since $D(q_j)=0$, (5.4.6) is given by

$$\lim_{q\to q_j}\frac{(q-q_j)}{D(q)} = \left\{\frac{d}{dq}D(q)\Big|_{q_j}\right\}^{-1} \ . \tag{5.4.7}$$

The latter term is from (5.2.23)

$$\frac{d}{dq}D(q) = \frac{q\,[1 + k_0 d|\mu_2|]\,[\mu_1 \sin(k_0\mu_1 d) - |\mu_2|\cos(k_0\mu_1 d)]}{\mu_1|\mu_2|} \; . \tag{5.4.8}$$

Use the fact that at these poles $D(q_j) = 0$ so that from (5.2.23) we have

$$\mu_{1j}\cos(k_0\mu_{1j}d) + |\mu_{2j}|\sin(k_0\mu_{1j}d) = 0 \; , \tag{5.4.9}$$

where

$$\mu_{1j} = (K_1^2 - q_j^2)^{1/2} \; , \tag{5.4.10}$$

and

$$|\mu_{2j}| = (q_j^2 - K_2^2)^{1/2} \; . \tag{5.4.11}$$

We can thus write the residues as

$$RES\Big[G_2^i(z, z; q_j^2)\Big] = U_j \sin(k_0\mu_{1j}z)\sin(k_0\mu_{1j}z) \; , \tag{5.4.12}$$

where

$$U_j = |\mu_{2j}|\,\Big[k_0 q_j\Big]^{-1}\Big[1 + k_0 d\,|\mu_{2j}|\Big]^{-1} \; , \tag{5.4.13}$$

and

$$RES\Big[G_2^e(z, z'; q_j^2)\Big] = L_j \sin(k_0\mu_{1j}z')e^{-k_0|\mu_{2j}|(z-d)} \; , \tag{5.4.14}$$

where

$$L_j = \mu_{1j}^2 U_j\Big[(K_1^2 - K_2^2)\sin(k_0\mu_{1j}d)\Big]^{-1} \; . \tag{5.4.15}$$

The residue terms in the exterior of the waveguide decay. The decay increases with frequency, mode number j and depth of penetration out of the waveguide. They essentially can be thought of as stripping the waveguide of energy.

We can combine (5.4.14), (5.4.12), and (5.4.13) to write the residue series representation in the interior of the waveguide. If we define the following functions

$$\phi_j(z) = \Big[\frac{2k_0|\mu_{2j}|}{1 + k_0 d|\mu_{2j}|}\Big]^{1/2}\sin(k_0\mu_{1j}z) \; , \tag{5.4.16}$$

we can write the representation in the symmetric form

$$RS^i = \frac{i}{4}\sum_{j=1}^{n} H_0^{(1)}(k_0 q_j r)\phi_j(z)\phi_j(z') \; . \tag{5.4.17}$$

The latter representation is the one often quoted in practice.

5.5 Fourier-Bessel Representation

Other representations are possible for the normal mode field. We list one which has the advantage of being solvable using Fourier transforms and has hence found its way into practice. The representation (5.4.1) can be written as

$$G(\boldsymbol{r}, \boldsymbol{r}') = \frac{k_0^2}{4\pi}\left\{\int_0^\infty H_0^{(1)}(k_0 qr) G_2(z, z'; q^2) q dq \right.$$

$$\left. + \int_{\infty e^{\pi i}}^0 H_0^{(1)}(k_0 qr) G_2(z, z'; q^2) q dq \right\} , \tag{5.5.1}$$

where we note that G_2 is an even function of q. In the second integral we use the transformation $q = q' \exp(\pi i)$, the symmetry condition

$$H_0^{(1)}\Big(k_0 r q' \exp(\pi i)\Big) = -H_0^{(2)}(k_0 r q') ,$$

and the Bessel function relation

$$H_0^{(1)}(x) + H_0^{(2)}(x) = 2J_0(x) .$$

We can then combine the two integrals in (5.5.1) to yield

$$G(\boldsymbol{r}, \boldsymbol{r}') = \frac{k_0^2}{2\pi}\int_0^\infty J_0(k_0 qr) G_2(z, z'; q^2) q dq , \tag{5.5.2}$$

which is referred to as *the Fourier-Bessel representation* of the solution. It forms the basis of what is referred to as the Fast Field Program (FFP) [5.5]. It can be solved using Fourier transforms as follows. First rewrite the representation as

$$G(\boldsymbol{r}, \boldsymbol{r}') = \int_0^\infty J_0(k_0 qr) F(q, z, z') q dq , \tag{5.5.3}$$

where

$$F(q, z, z') = \frac{k_0^2}{2\pi} G_2(z, z'; q^2) . \tag{5.5.4}$$

Introduce the transformation $q = e^{-\tau}$ (called a log stretch transform) and define

$$F_1(\tau, z, z') = F(e^{-\tau}, z, z') . \tag{5.5.5}$$

Then (5.5.3) can be rewritten as

$$G(r, z, z') = \int_{-\infty}^\infty J_0(k_0 e^{-\tau} r) F_1(\tau, z, z') e^{-2\tau} d\tau . \tag{5.5.6}$$

Introduce the second transformation $r = e^\rho$ and define

$$g(\rho, z, z') = G(e^\rho, z, z') \ , \tag{5.5.7}$$

and

$$F_2(\tau, z, z') = F_1(\tau, z, z')e^{-2\tau} \ . \tag{5.5.8}$$

Then (5.5.6) can be written as a convolution

$$g(\rho, z, z') = \int_{-\infty}^{\infty} K(\rho - \tau)F_2(\tau, z, z')d\tau \ , \tag{5.5.9}$$

where

$$K(\rho - \tau) = J_0(k_0\, e^{\rho-\tau}) \ . \tag{5.5.10}$$

If we define the Fourier transform of g

$$\tilde{g}(\alpha, z, z') = \int_{-\infty}^{\infty} e^{-i\alpha\rho} g(\rho, z, z)d\rho \ , \tag{5.5.11}$$

then the transform of (5.5.9) is given by

$$\tilde{g}(\alpha, z, z') = \widetilde{K}(\alpha)\tilde{F}_2(\alpha, z, z') \ , \tag{5.5.12}$$

in terms of the corresponding Fourier transforms of K and F_2. These latter transforms can also be written conveniently as Mellin transforms. We have

$$\begin{aligned} \widetilde{K}(\alpha) &= \int_{-\infty}^{\infty} e^{-i\alpha\sigma} K(\sigma)d\sigma \\ &= \int_{-\infty}^{\infty} e^{-i\alpha\sigma} J_0(k_0 e^\sigma)d\sigma \ ; \quad \sigma = \ln\mu \\ &= \int_0^{\infty} \mu^{-1-i\alpha} J_0(k_0\mu)d\mu \ , \end{aligned} \tag{5.5.13}$$

and

$$\begin{aligned} \widetilde{F}_2(\alpha, z, z') &= \int_{-\infty}^{\infty} e^{-i\alpha\tau} F_2(\tau, z, z')d\tau \\ &= \int_{-\infty}^{\infty} e^{-i\alpha\tau} F(e^{-\tau}, z, z')e^{-2\tau}d\tau \ \ ; \tau = \ln\nu \\ &= \int_0^{\infty} \nu^{-3-i\alpha} F(\nu^{-1}, z, z')d\nu ; \quad \xi = \nu^{-1} \\ &= \int_0^{\infty} \xi^{i\alpha+1} F(\xi, z, z)d\xi \ . \end{aligned} \tag{5.5.14}$$

Both these relations are also useful in practice.

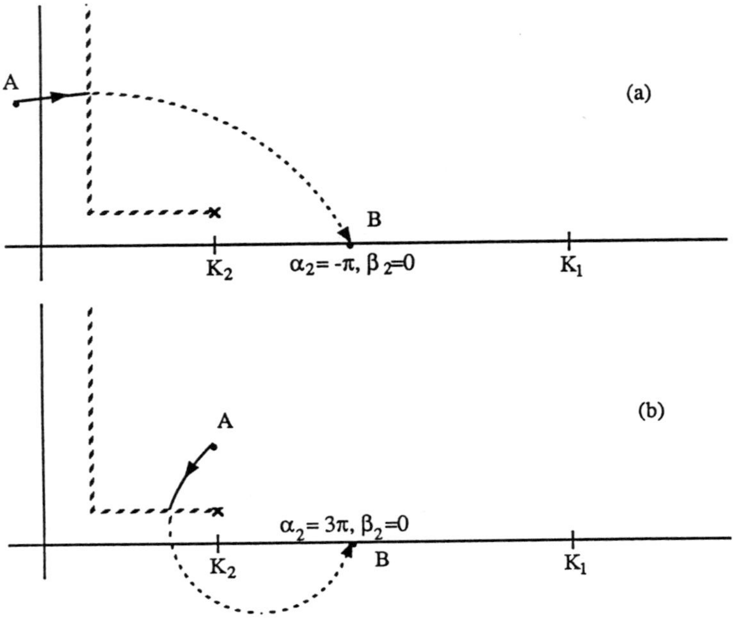

Fig. 5.8. Two possible analytic continuations from a first sheet point A to a point B on the second sheet. In (a) the second sheet is defined by continuation through the lower lip of the K_2 branch cut, and in (b) it is defined by continuation through the upper lip of the cut. Solid lines on the path are postions on the first sheet, dashed lines on the second sheet.

5.6 Pseudo-proper Modes

We found the proper modes by finding the zeroes of the denominator $D(q)$ on the first or proper sheet. The *pseudo-proper modes* are the zeroes of $D(q)$ which occur on the second sheet. They occur at almost the same position as the proper modes and are often referred to as the *shadow poles* of the proper modes. They are real solutions. We can write $D(q) = 0$ from (5.2.23) as

$$\cot(k_0\mu_1 d) = i\mu_1^{-1}|\mu_2|\left\{\cos\left[\frac{1}{2}(\alpha_2+\beta_2)\right]\right.$$
$$\left. + i\sin\left[\frac{1}{2}(\alpha_2+\beta_2)\right]\right\} , \qquad (5.6.1)$$

where α_2 and β_2 are the angles of the square roots in μ_2. We can define the second sheet in several ways, both of which involve a continuation through the branch cut at K_2. The two continuations to the region where $K_2 < q_r < K_1$ on the second sheet are illustrated in Fig. 5.8.

If we choose the second sheet as illustrated in Fig.(5.8b), then $\alpha_2 + \beta_2 = 3\pi$ in the region $K_2 < q_r < K_1$, and (5.6.1) becomes

$$\cot(k_0\mu_1 d) = |\mu_2|/\mu_1 \ . \tag{5.6.2}$$

This is similar to (5.3.28) except for the sign change on the *rhs*. The intersections can be seen to occur using the illustration in Fig.(5.5) but reflecting the lower curve through the real axis. The intersections do not occur at the same positions as the q_j. In fact, as we sketched the example, only two shadow poles occur. Note that the same result would occur had we chosen the second sheet as in Fig.(5.8a).

On the second sheet, however, $\mu_{2i} < 0$ so that the exterior Green's function (5.2.26) would behave as a growing exponential as $z \to \infty$. These are unphysical solutions and are thus rejected. Note also that the pseudo-proper modes are "far" from the physical region (first sheet). Here "far" means the distance required to analytically continue from a point on the physical sheet to the position of the pseudo-proper modes.

5.7 Leaky Waves

We showed that the only poles which occurred on the first sheet were proper modes. In particular, no other complex poles occurred on this sheet. This is not true on the second sheet. The *leaky waves* are complex poles which occur on the second sheet. We derive their approximate position and point out why they can play a role in the physical solution. (In the Stickler-Ammicht method [5.1, 5.9], both shadow poles and leaky waves are included.) Set $D(q) = 0$ from (5.2.23) and write the result in exponential form as

$$e^{2ik_0 d\mu_1} = \frac{\mu_2 + \mu_1}{\mu_2 - \mu_1} \ . \tag{5.7.1}$$

This equation can have periodic solutions of the form

$$2k_0 d\mu_1 = -i\ln\left(\frac{\mu_2 + \mu_1}{\mu_2 - \mu_1}\right) + 2n\pi \ , \tag{5.7.2}$$

where n is an integer ($n = 1, 2, 3, \cdots$). (We exclude $n = 0$ since this is the first sheet.) We can easily find the approximate position of these poles for large values of q. Let

$$q = i\tau, \quad \overline{\tau} = \pm\tau \ , \tag{5.7.3}$$

where we choose $\overline{\tau}$ in anticipation of the fact that (5.7.2) is an even function of q. We can thus write

$$\begin{aligned} \mu_1 &= \left(K_1^2 + \tau^2\right)^{1/2} \\ &= \overline{\tau}\left(1 + K_1^2/\tau^2\right)^{1/2} \\ &\approx \overline{\tau}\left(1 + \frac{1}{2}\frac{K_1^2}{\tau^2}\right) \end{aligned} \tag{5.7.4}$$

where the approximation is true for large τ. Similarly

$$\mu_2 \approx \overline{\tau}\left(1 + \frac{1}{2}\frac{K_2^2}{\tau^2}\right) \ . \tag{5.7.5}$$

Substituting the results in (5.7.2) we get

$$2k_0 d\overline{\tau}\left(1 + \frac{1}{2}\frac{K_1^2}{\tau^2}\right) \approx -i\ln\left\{\frac{2 + \frac{1}{2}\frac{(K_2^2 + K_1^2)}{\tau^2}}{\frac{1}{2}\frac{K_2^2 - K_1^2}{\tau^2}}\right\} + 2n\pi \ , \tag{5.7.6}$$

The numerator of the ln term has the form $\ln(1+x) \approx x$ where x is small, and the denominator $\ln y \approx -1 + y$ where y is small. Expanding the logarithm in this way we can write (5.7.6) as

$$\begin{aligned} 2k_0 d\overline{\tau}\left(1 + \frac{1}{2}\frac{K_1^2}{\tau^2}\right) &\approx -i\left(\frac{K_2^2 + K_1^2}{4\tau^2}\right) \\ &+ i\left(-1 + \pi i + \frac{K_1^2 - K_2^2}{4\tau^2}\right) + 2n\pi \ . \end{aligned} \tag{5.7.7}$$

This is a cubic equation in τ, but in lowest order can be approximately written as (for large τ and large n)

$$\overline{\tau} \sim \frac{-i}{2k_0 d} + \frac{n\pi}{k_0 d} \ . \tag{5.7.8}$$

Using $\overline{q} = i\overline{\tau} = \pm q$ we can write the solutions of (5.7.8) as

$$q_n = \pm\frac{1}{2k_0 d} \pm i\frac{n\pi}{k_0 d} \ . \tag{5.7.9}$$

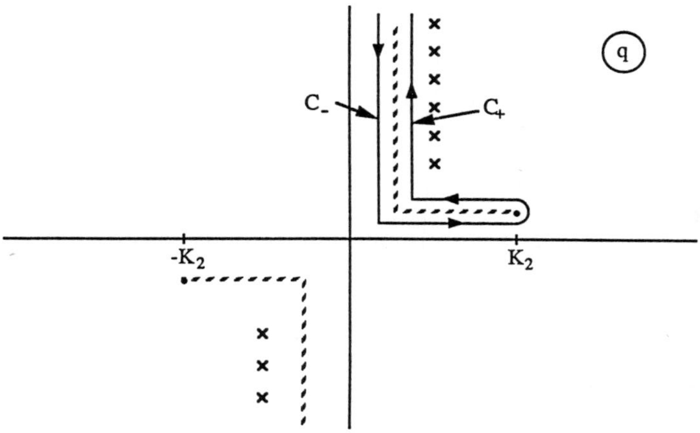

Fig. 5.9. Approximate asymptotic positions of the leaky wave poles ($\times$) in the first and third quadrants of the second sheet of the $q-$ plane. The BLI is also illustrated.

The q_n values thus occur in the first and third quadrants. They are schematically illustrated in Fig.(5.9). They can be easily seen to lie on the second sheet using an argument analogous to that in Chap. 4 to find the orientation of the branch cuts. Write (5.7.9) as

$$q_n = a + ib_n \quad , \tag{5.7.10}$$

where $ab_n > 0$. Then at these points

$$\mu_2 = \left(K_2^2 - q_n^2\right)^{1/2} = \mu_{2r} + i\mu_{2i} \quad . \tag{5.7.11}$$

Squaring (5.7.11), substituting q_n from (5.7.10), and equating the real and imaginary parts, we get for the imaginary part

$$\mu_{2r}\mu_{2i} = -ab_n \quad . \tag{5.7.12}$$

In the first and third quadrants $\mu_{2r} > 0$, and $\mu_{2i} < 0$ on the entire second sheet. Since $ab_n > 0$ these solutions occur on the second sheet.

For physical reasons we have excluded the poles in the third quadrant since $q_i < 0$ and we would get growing solutions in range. The poles in the first quadrant however are physically interesting. The reason is that they occur close (in the sense of analytic continuation) to the C_- contour integral in Fig.(5.9). They are not close to C_+. This corresponds to a definition of the second sheet as illustrated in

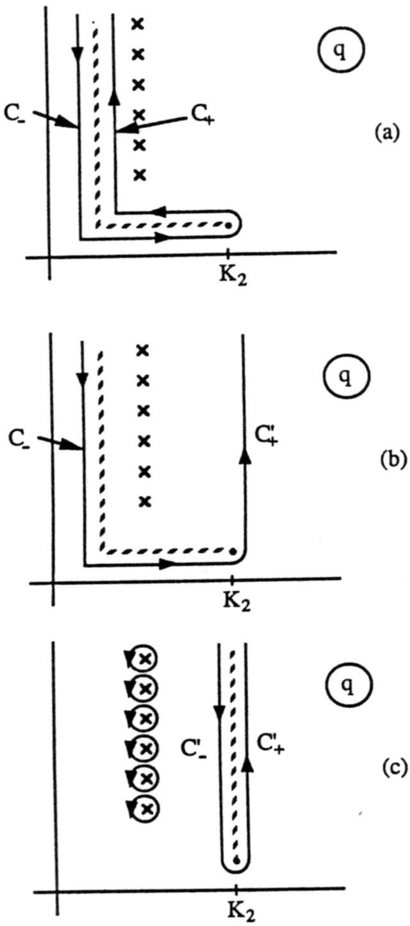

Fig. 5.10. (a) The BLI due to Ewing, Jardetzky and Press (EJP). The leaky poles (x) are on the second sheet. (b) Distortion of the C contour to C'. The leaky poles are still on the second sheet. (c) The Pekeris (PK) version of the BLI where the branch cut is vertical. The leaky poles now contribute a residue series and appear on the first sheet of this new cut.

Fig.(5.8a). The BLI we have described is due to Ewing, Jardetsky and Press (EJP, [5.6]). Since we have proved that its integrand is analytic in the first quadrant of the first sheet, C_+ can be distorted to run vertically as C'_+ illustrated in Fig.(5.10b). The branch cut is distorted to run vertically as it was in the original work of Pekeris (PK, [5.8]) and C_- is distorted to follow it as illustrated in Fig.(5.10c). The leaky poles (whose position is fixed) now appear on the first sheet as a residue sum. (Strictly speaking, the EJP version of the modal expansion is uniform in r and z, whereas the P version is not due to the asymptotic part of the contour difference between the two versions. The latter does not give a zero contribution for fixed r as $z \to \infty$. This is mathematically interesting but has no physical consequences.)

We illustrate the result schematically as

$$BLI = \int_{EJP} = \int_{PK} + LS \ , \tag{5.7.13}$$

where the integral with subscript PK is the Pekeris BLI evaluated in the next section. The symbol LS represents the residue series due to the leaky poles or *leaky sum.* The latter can be evaluated as a residue series in a manner analogous to that used to evaluate the proper modes. We have that

$$LS = \frac{ik_0^2}{2} \sum_{n=1}^{\infty} H_0^{(1)}(k_0 q_n r)\, q_n \; RESLP\Big[G_2(z, z'; q_n^2)\Big] \ , \tag{5.7.14}$$

where $RESLP$ stands for the residue of G_2 at a leaky pole, where each pole must be determined from (5.7.1) and each is given asymptotically by (5.7.9) with the + signs. We calculate the residues as in Sect. 5.4. From (5.2.24) we have that in the interior of the waveguide

$$RESLP\Big[G_2^i(z, z'; q_n^2)\Big] = \frac{\sin(k_0 \mu_{1n} z_<) N_1(q_n)}{k_0 \mu_{1n} \left[dD/dq\right]_{q_n}} \ , \tag{5.7.15}$$

and for the exterior we get from (5.2.26)

$$RESLP\Big[G_2^e(z, z'; q_n^2)\Big] = \frac{\sin(k_0 \mu_{1n} z') e^{ik_0 \mu_{2n}(z-d)}}{K_0 \left[dD/dq\right]_{qn}} \ . \tag{5.7.16}$$

Using (5.2.23) and (5.2.25) we can calculate these residue terms using

$$\frac{dD}{dq} = \frac{q}{\mu_1 \mu_2}\Big[\mu_2 k_0 d + i\Big]\Big[\mu_1 \sin(k_0 \mu_1 d) + i\mu_2 \cos(k_0 \mu_1 d)\Big] \ . \tag{5.7.17}$$

Using $D(q_n) = 0$ this can be written as

$$\frac{dD}{dq}(q_n) = \frac{q_n(K_1^2 - K_2^2)(\mu_{2n}k_0 d + i)\sin(k_0\mu_{1n}d)}{\mu_{1n}^2\mu_{2n}} \ . \qquad (5.7.18)$$

We also find that

$$\frac{N_1(q_n)}{\left[dD/dq\right]_{qn}} = \frac{\mu_{1n}\mu_{2n}\sin(k_0\mu_{1n}z_>)}{q_n\left[\mu_{2n}k_0 d + i\right]} \ . \qquad (5.7.19)$$

The leaky sum in the interior of the waveguide can thus be written using (5.7.14), (5.7.15) and (5.7.19) as

$$LS^i = \frac{i}{4}\sum_{n=1}^{\infty} H_0^{(1)}\left(k_0 q_n r\right)\psi_n(z)\psi_n(z') \ , \qquad (5.7.20)$$

where we have defined the leaky-eigenfunctions

$$\psi_n(z) = \left\{\frac{2k_0\mu_{2n}}{\mu_{2n}k_0 d + i}\right\}^{1/2} \sin(k_0\mu_{1n}z) \ . \qquad (5.7.21)$$

This leaky sum appears to have the same form as the residue series (5.4.17), but here both μ_{1n} and μ_{2n} as well as q_n are all complex, with $\mathrm{Im}q_n$ positive. Asymptotically, for r large, this leads to waves which decay in r both algebraically as $r^{-1/2}$ from the Hankel function and, more importantly, exponentially as $\exp[-k_0 r \mathrm{Im} q_n]$. The larger the value of n, the faster the decay. These leaky waves can only play a role in the solution at short range, and then only the lower order leaky waves. Since μ_{in} is complex, $(i = 1, 2)$, the leaky-eigenfunctions behave as both growing and decaying exponentials by (5.7.21). This is still physically acceptable since they do so in the bounded region of the waveguide. Since μ_{2n} has a negative imaginary part, we see from (5.7.16) that the exterior solution grows exponentially in depth z. In order for any solution to remain valid it must converge. We can see from (5.7.20) and (5.7.21) that the interior solution converges provided

$$- r\mathrm{Im}\, q_n + (z + z')\mathrm{Im}\, \mu_{1n} < 0 \ . \qquad (5.7.22)$$

This follows from the asymptotic evaluation of the Hankel function, and the combination of its phase with the growing depth-dependent exponentials of the sines. From (5.7.14) and (5.7.16), the exterior solution converges provided

$$- r\text{Im}\, q_n + z|\text{Im}\mu_{2n}| + z'\text{Im}\,\mu_{1n} < 0 \ . \tag{5.7.23}$$

Basically, the leaky wave sum will converge provided we are far enough out in range. Then, however, it is of course exponentially decaying.

5.8 Branch Line Integral (BLI)

In this section we asymptotically evaluate the *BLI* for the interior of the waveguide. From (5.4.3) we have that

$$BLI = \frac{k_0^2}{4\pi} \int_{C_-} H_0^{(1)}(k_0 qr) \Delta_2 G_2^i(z, z'; q^2) q dq \ , \tag{5.8.1}$$

where the discontinuity can be written from (5.3.18) as

$$\Delta_2 G_2(z, z'; q^2) = |\mu_2| A(q) \ , \tag{5.8.2}$$

where

$$A(q) = \frac{2i}{k_0} \frac{\sin(k_0\mu_1 z_<)\sin(k_0\mu_1 z_>)}{\mu_1^2\cos^2(k_0\mu_1 d) + |\mu_2|^2\sin^2(k_0\mu_1 d)} \ , \tag{5.8.3}$$

is an analytic function in the neighborhood of the branch point $q = K_2$. The contour C_- runs from $i\infty$ to K_2 and is illustrated in Fig.(5.7). Using asymptotics on the Hankel function we have

$$BLI \sim \frac{k_0^2}{4\pi}\left(\frac{2}{\pi k_0 r}\right)^{1/2} e^{-i\pi/4}$$

$$\cdot \int_{i\infty}^{K_2} e^{ik_0 qr} \left|K_2^2 - q^2\right|^{1/2} A(q) q^{1/2} dq \ . \tag{5.8.4}$$

We remarked in the previous section that our BLI was due to EJP. Shifting it to the PK contour we can write the integral in (5.8.4) as

$$\int_{i\infty}^{K_2} = \int_{i\infty + K_2}^{K_2} + LS \ . \tag{5.8.5}$$

This is illustrated in Fig.(5.10). We evaluated the leaky pole sum in Sect. 5.7. Here we evaluate the line integral due to Pekeris which is

$$PK \sim \frac{k_0^2}{4\pi}\left(\frac{2}{\pi k_0 r}\right)^{1/2i} e^{-i\pi/4}$$

$$\cdot \int_{i\infty + K_2}^{K_2} e^{ik_0 qr} \left|K_2^2 - q^2\right|^{1/2} A(q) q^{1/2} dq \ . \tag{5.8.6}$$

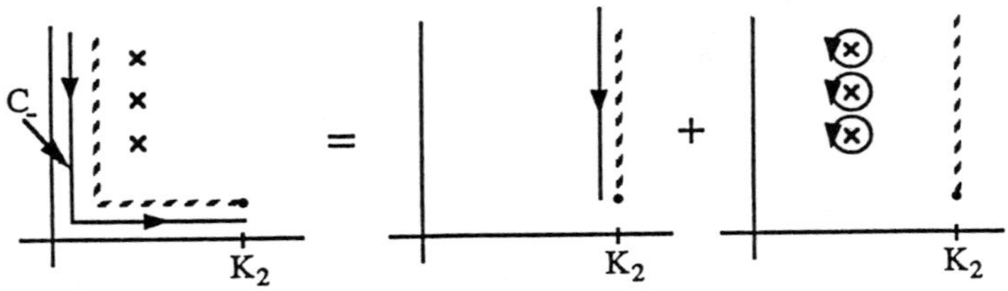

Fig. 5.11. Illustration of (5.8.5). The *EJP* line integral equals the *PK* line integral plus the leaky poles.

Change variables in the integral in (5.8.6). Let $q = y + K_2$ (shift to the branch point) so that it becomes

$$I = \int_{i\infty}^{0} e^{ik_0 r(y+K_2)} \left|-y(y+2K_2)\right|^{1/2} A(y+K_2)(y+K_2)^{1/2} dy \ . \quad (5.8.7)$$

Let $y = i\tau$ so that

$$I = i \int_{\infty}^{0} e^{-k_0 r\tau} e^{ik_0 rK_2} \tau^{1/2} \left|2K_2 + i\tau\right|^{1/2} \cdot A(K_2 + i\tau) \left[K_2 + i\tau\right]^{1/2} d\tau \ . \quad (5.8.8)$$

Next expand the analytic part of the integrand about $\tau = 0$ and keep only the leading order term

$$I \approx -i\sqrt{2} K_2 A(K_2) e^{ik_0 rK_2} \int_0^{\infty} e^{-k_0 r\tau} \tau^{1/2} d\tau \ . \quad (5.8.9)$$

Finally, let $\sigma = k_0 r\tau$ in the integrand so that

$$I \approx -\frac{i\sqrt{2}\, K_2 A(K_2) e^{ik_0 rK_2}}{(k_0 r)^{3/2}} \int_0^{\infty} e^{-\sigma} \sigma^{1/2} d\sigma \ . \quad (5.8.10)$$

The integral in (5.8.10) is just equal to $\sqrt{\pi}/2$ so that (5.8.6) becomes

$$PK \sim \frac{K_2 A(K_2)}{4\pi r^2} e^{ik_0 rK_2} e^{-3\pi i/4} \ . \quad (5.8.11)$$

Since $A(k)$ behaves like k_0^{-1} or ω^{-1}, the critical behavior of the line integral is

$$PK \sim 0\left(\omega^{-1} r^{-2}\right) \ . \quad (5.8.12)$$

Note that it does not decay exponentially. The real exponential in range is an indication that the wave has been modified by the parameters of the medium. Since $k_0 K = \omega/c_2$ the wave exists in region 2.

5.9 Virtual Modes

Virtual modes are defined as the *resonance-like enhancements* of the integrand of the *BLI* when the denominator is at a minimum. They are also referred to as *continuous modes* (because they occur along the *BLI* which is the continuous spectrum of the problem) or *improper modes.* The essential reason they occur is because of the nearby presence of the leaky poles. From (5.4.3) we can write

$$BLI = \frac{k_0^2}{4\pi} \int_{C_-} H_0^{(1)}(k_0 rq) \Delta_2 G_2(z, z'; q^2) q dq \ , \tag{5.9.1}$$

where the contour of integration C_- follows the path from $i\infty$ to K_2 (see Fig.5.7). We change the integration variable to $\mu_2 = (K_2^2 - q^2)^{1/2}$ which follows the path from ∞ to 0. If in addition we choose to evaluate the *BLI* in the waveguide we can write using (5.3.18)

$$\Delta_2 G_2^i(z, z'; q^2) = 2i\mu_2 k_0^{-1} g(z, z'; \mu_2^2) \ , \tag{5.9.2}$$

where

$$g(z, z'; \mu_2^2) = \sin(k_0 z' \mu_1) \sin(k_0 z \mu_1) D_2^{-1}(q) \ , \tag{5.9.3}$$

and

$$D_2(q) = \mu_1^2 \cos^2(k_0 d\mu_1) + \mu_2^2 \sin^2(k_0 d\mu_1) \ , \tag{5.9.4}$$

is just (5.3.17). Here we use the fact that

$$\mu_1^2 = W + \mu_2^2 ; \quad W \doteq K_1^2 - K_2^2 > 0 \ . \tag{5.9.5}$$

Combining these results, we can write (5.9.1) as

$$BLI = \frac{ik_0}{2\pi} \int_0^\infty H_0^{(1)} \left(k_0 r (K_2^2 - \mu_2^2)^{1/2} \right) g(z, z'; \mu_2^2) \mu_2^2 \, d\mu_2 \ . \tag{5.9.6}$$

This is the integral we work with in the following analysis.

We first establish some properties of the denominator D_2. Using (5.9.5) we can rewrite (5.9.4) as

$$D_2(q) = \mu_2^2 + W \cos^2(k_0 d\mu_1) \ . \tag{5.9.7}$$

The first derivative of (5.9.7) is given by

$$\frac{dD_2}{d\mu_2} = \mu_2 \left\{ 2 - k_0 d W \mu_1^{-1} \sin(2k_0 d\mu_1) \right\} \ , \tag{5.9.8}$$

and its second derivative can be written using (5.9.8) as

$$\frac{d^2D_2}{d\mu_2^2} = \frac{1}{\mu_2}\frac{dD_2}{d\mu_2}\left(1 - \frac{\mu_2^2}{\mu_1^2}\right) + 2\frac{\mu_2^2}{\mu_1^2}\left\{1 - (k_0d)^2 W\cos(2k_0d\mu_1)\right\} \ . \quad (5.9.9)$$

There are several approaches to the minimization of $D_2(q)$ and we present them next.

Definition 1: This method is related to that of Tindle, Stamp, and Guthrie (TSG, [5.10]) and others (see [5.4]). Consider (5.9.4). It consists of two positive terms. Since $\mu_1^2 > \mu_2^2$ here, we minimize this positive quantity by choosing

$$\cos\left[k_0d\mu_1\right] = 0 \ , \quad (5.9.10)$$

which results in $D_2(q)$ having periodic relative minima occurring at

$$\mu_{1m} = (m - 1/2)(\pi/k_0d)\,; \ \ m = 0, \pm 1, \pm 2, \cdots \ , \quad (5.9.11)$$

or at values q_m

$$q_m = +\left[K_1^2 - \mu_{1m}^2\right]^{1/2} . \quad (5.9.12)$$

We require $0 < q_m < K_2$ if real and $q_m = i|q_m|$ if on the imaginary axis. The resulting maximum values of the integrand of (5.9.1) can be found using (5.9.5) to occur at

$$\mu_{2m} = \left[\mu_{1m}^2 - W\right]^{1/2} \ . \quad (5.9.13)$$

In the neighborhood of these resonance-like terms, approximate the denominator using a quadratic dependence

$$D_2(q) \cong D_2(q_m)\left\{1 + (\mu_2 - \mu_{2m})^2\alpha_m\right\} \ , \quad (5.9.14)$$

where by (5.9.4)

$$D_2(q_m) = \mu_{2m}^2 \ , \quad (5.9.15)$$

and with α_m defined by

$$\alpha_m = \left[2D_2(q_m)\right]^{-1}\left.\frac{d^2D_2}{d\mu_2^2}\right|_m \ . \quad (5.9.16)$$

At these points we can approximately write

$$g(z, z'; \mu_2^2) \cong \sum_m \frac{R_m}{1 + \left[\mu_2 - \mu_{2m}\right]^2\alpha_m} \ , \quad (5.9.17)$$

where the residues are defined as

$$R_m = \sin\left[k_0 z' \mu_{1m}\right] \sin\left[k_0 z \mu_{1m}\right] \left[D_2(q_m)\right]^{-1} \quad . \tag{5.9.18}$$

Note that, strictly speaking, g is a function of μ_2^2 whereas the *rhs* of (5.9.17) is not, so that (5.9.17) is only an approximation. Using these approximations the *BLI* from (5.9.6) becomes

$$BLI \cong \frac{ik_0}{2\pi} \sum_m R_m \mu_{2m}^2 \int_0^\infty \frac{H_0^{(1)}\left(k_0 r[K_2^2 - \mu_2^2]^{1/2}\right)}{1 + [\mu_2 - \mu_{2m}]^2 \alpha_m} d\mu_2 \ , \tag{5.9.19}$$

where we have expanded the μ_2 in the integrand about μ_{2m}. For each m, the integral in (5.9.19) is really over a small region Δ_m about the resonance position μ_{2m}.

For small $k_0 r$ the integrals in (5.9.19) are best evaluated numerically. This can be done since the oscillations in the Hankel function are not rapid. For $k_0 r$ large we can estimate the behavior of the integrals using asymptotics on the Hankel function

$$H_0^{(1)}\left(k_0 r[K_2^2 - \mu_2^2]^{1/2}\right) \sim \left\{\frac{-2i}{\pi k_0 r\left[K_2^2 - \mu_2^2\right]}\right\}^{1/2} e^{ik_0 r[K_2^2 - \mu_2^2]^{1/2}} \ , \tag{5.9.20}$$

and an expansion of the square root

$$\left[K_2^2 - \mu_2^2\right]^{1/2} \approx \left[K_2^2 - \mu_{2m}^2\right]^{1/2} + [\mu_2 - \mu_{2m}]\left\{-\mu_{2m} \Big/ \left[K_2^2 - \mu_{2m}^2\right]^{1/2}\right\}$$

$$= q_m - [\mu_{2m}/q_m]\,[\mu_2 - \mu_{2m}] \quad . \tag{5.9.21}$$

The result yields for (5.9.19)

$$BLI \sim \left(\frac{ik_0}{2\pi^3 r}\right)^{1/2} \sum_m R_m \mu_{2m}^2 q_m^{-1/2} e^{ik_0 r q_m} I_m \ , \tag{5.9.22}$$

where the integral are defined as

$$I_m = \int_0^\infty \frac{e^{-iL_m(\mu_2 - \mu_{2m})}}{1 + [\mu_2 - \mu_{2m}]^2 \alpha_m} d\mu_2 \ , \tag{5.9.23}$$

with

$$L_m = k_0 r \mu_{2m}/q_m \quad . \tag{5.9.24}$$

There are two possible evaluations of (5.9.23), the first for real values of q_m, and the second for imaginary q_m.

Case 1: q_m real, $0 < q_m < K_2$ This is the case of *real virtual modes.* Since $L_m > 0$, the integrand in (5.9.23) rapidly vanishes as $\mu_2 \to \pm\infty$ for fixed $k_0 r$ and fixed m. Introduce a change of variables

$$\varepsilon = \mu_2 - \mu_{2m} \ , \tag{5.9.25}$$

so that (5.9.23) becomes

$$I_m = \int_{-\mu_{2m}}^{\infty} \frac{e^{-iL_m\varepsilon}}{1+\varepsilon^2\alpha_m} d\varepsilon \ . \tag{5.9.26}$$

We approximate the integral by extending the integration to $-\infty$ and rewrite it as

$$I_m \approx \frac{1}{\alpha_m}\int_{-\infty}^{\infty} \frac{e^{-iL_m\varepsilon}}{(\varepsilon + i\alpha_m^{-1/2})(\varepsilon - i\alpha_m^{-1/2})} d\varepsilon \ . \tag{5.9.27}$$

Next, evaluate α_m from (5.9.16) using (5.9.15). Using $\sin 2x = 2\sin x\cos x$ it follows from (5.9.8) that

$$\left.\frac{dD_2}{d\mu_2}\right|_m = 2\mu_{2m} \ . \tag{5.9.28}$$

Further using $\cos 2x = \cos^2 x - \sin^2 x$ and (5.9.10) it follows from (5.9.9) that

$$\left.\frac{d^2D_2}{d\mu_2^2}\right|_m = 2\left\{1 + \mu_{2m}^2\mu_{1m}^{-2}(k_0 d)^2 W\right\} \ . \tag{5.9.29}$$

It therefore follows from (5.9.16) that

$$\alpha_m = \mu_{2m}^{-2} + \mu_{1m}^{-2}(k_0 d)^2 W \ , \tag{5.9.30}$$

which is a positive real quantity. Evaluation of (5.9.27) using residue calculus methods (close the contour in the lower half plane) yields

$$I_m \cong \pi\alpha_m^{-1/2} e^{-k_0 r\mu_{2m}/q_m\alpha_m^{1/2}} \ . \tag{5.9.31}$$

Thus, for real virtual modes, (5.9.22) and (5.9.31) yield

$$BLI \sim \left(\frac{ik_0}{2r}\right)^{1/2} \sum_m R_m\mu_{2m}^2(q_m\alpha_m)^{-1/2} e^{k_0 r\left[iq_m - \mu_{2m}/q_m\alpha_m^{1/2}\right]} \ . \tag{5.9.32}$$

Each mode thus exponentially decays in range (at long range). This is as expected since

$$q_m = \frac{c_0}{c_1}\sin\theta_m < K_2 = \frac{c_0}{c_2} \ , \tag{5.9.33}$$

and thus

$$\sin\theta_m < \frac{c_1}{c_2} = \sin\theta_c \ . \tag{5.9.34}$$

It follows that $\theta_m < \theta_c$, i.e. the equivalent mode angle is less than critical. This results in a transmitted field or equivalently a leakage of energy out of the duct. Note that for high frequency, the decay is frequency independent since $\alpha_m \sim (k_0 d)^2$.

We give a brief calulation on the number of *real virtual modes* which can exist. These are modes where

$$0 < q_m < K_2 \ . \tag{5.9.35}$$

Since $q_m = (K_1^2 - \mu_{1m}^2)^{1/2}$ we can manipulate these inequalities to yield

$$W^{1/2} < \mu_{1m} < K_1 \ , \tag{5.9.36}$$

where W is defined in (5.9.5). Using (5.9.11) we thus get

$$W^{1/2} k_0 d/\pi < m - 1/2 < k_0 d K_1/\pi \ , \tag{5.9.37}$$

for the number of possible positive integers m which form real virtual modes. Using $K_j = c_0/c_j$ and $k_0 = 2\pi f/c_0$ where f is frequency this can also be written as

$$2fd(c_1^{-2} - c_2^{-2})^{1/2} < m - 1/2 < 2fd/c_1 \ . \tag{5.9.38}$$

As a simple example, consider a shallow waveguide ($d = 300m$) for a frequency $f = 100Hz$ with $c_0 = 1500m/s$, $K_1 = 1$ and $K_2 = 0.9$. Then $k_0 d/\pi = 40$ and

$$40(.19)^{1/2} < m - 1/2 < 40 \ . \tag{5.9.39}$$

Solutions exist for $m = 18, \cdots, 40$ for a total of 23 real virtual modes. The thicker the waveguide, the higher the frequency, or the greater the bottom sound speed contrast all yield more virtual modes. Their importance can arise in several situations in spite of the fact that each mode exponentially decays. (Whether the BLI sum decays is an open question.) The situations are first of all, of course, the near field close to the source since they haven't died away yet. Secondly, at low frequencies, since in this situation there may be few or no proper modes. Finally, in surface ducts where the sound speed increases from the surface to the bottom (of the duct), generally all modes are continuous.

Case 2: $q_m = i|q_m|$, $0 < |q_m| < \infty$ This is the case of *imaginary virtual modes*. For this case $\mu_{2m} = \{K_2^2 + |q_m|^2\}^{1/2}$ so that we have $\mu_{2m} : K_2 \to \infty$. For these modes we can write the I_m integrals from (5.9.23) as

$$I_m \approx \int_{K_2}^{\infty} \frac{e^{-ik_0 r(\mu_2 - \mu_{2m})\mu_{2m}/q_m}}{1 + [\mu_2 - \mu_{2m}]^2 \alpha_m} d\mu_2 \ . \tag{5.9.40}$$

If we let $\varepsilon = \mu_2 - \mu_{2m}$, $q_m = i|q_m|$ this becomes

$$I_m \approx \int_{K_2 - \mu_{2m}}^{\infty} \frac{e^{-|L_m|\varepsilon}}{1 + \varepsilon^2 \alpha_m} d\varepsilon \ , \tag{5.9.41}$$

where L_m is defined in (5.9.24). Since $\alpha_m > 0$, this integral can be bounded

$$\begin{aligned} I_m &< \int_{K_2 - \mu_{2m}}^{\infty} e^{-|L_m|\varepsilon} d\varepsilon \\ &= \frac{|q_m|}{k_0 r \mu_{2m}} e^{-k_0 r \mu_{2m}[K_2 - \mu_{2m}]/|q_m|} \ . \end{aligned} \tag{5.9.42}$$

Combining this with the exponential factor in (5.9.22) we can write the latter as

$$\exp\left\{-k_0 r \left[K_2 \mu_{2m} - K_2^2\right] / |q_m|\right\} \ . \tag{5.9.43}$$

For m large (which it can be since we could have many real virtual modes), the first term in square brackets in (5.9.43) predominates. Using this result and (5.9.42) in (5.9.22) we see that for imaginary virtual modes we can write

$$BLI \sim \left(\frac{1}{2\pi^3}\right)^{1/2} \frac{1}{k_0^{1/2} r^{3/2}} \sum_m R_m \mu_{2m} |q_m|^{1/2} e^{-k_0 r K_2 \mu_{2m}/|q_m|} \ . \tag{5.9.44}$$

Both the amplitude decay and the exponential modal decay differ from those of the real virtual modes in (5.9.32). The overall amplitude decays with a factor $(k_0 r)^{-1}$ faster, but the exponential decay of each mode is actually slower than for each real virtual mode.

Definition 2: The TSG method produced relative minima in the denominator. This method produces a strict denominator minimization by setting

$$\frac{dD_2}{d\mu_2} = 0; \quad \frac{d^2 D_2}{d\mu_2^2} > 0 \ . \tag{5.9.45}$$

A Taylor expansion of D_2 about the resulting μ_{2m} or q_m yields

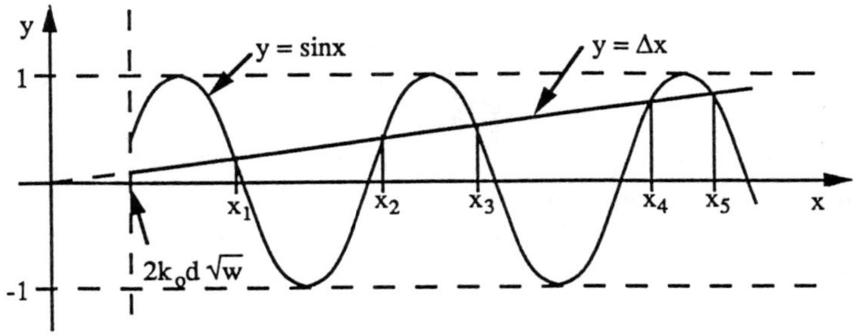

Fig. 5.12. Illustration of the eigenvalue determination for the denominator minimized virtual modes. Five possible roots x_m are illustrated. Only some of these correspond to true minimum denominator positions.

$$D_2(q) \cong D_2(q_m)+(\mu_2-\mu_{2m})\left.\frac{dD_2}{d\mu_2}\right|_m+\frac{1}{2}(\mu_2-\mu_{2m})^2\left.\frac{d^2D_2}{d\mu_2^2}\right|_m \ . \quad (5.9.46)$$

Since the first derivative vanishes (this forms the eigenvalue condition below) we can write

$$D_2(q) \cong D_2(q_m)\left\{1+(\mu_2-\mu_{2m})^2\alpha_m\right\} \ , \quad (5.9.47)$$

where

$$\alpha_m = [2D_2(q_m)]^{-1}\left.\frac{d^2D_2}{d\mu_2^2}\right|_m \ . \quad (5.9.48)$$

This is, of course, the same type of expansion we assumed in *TSG*, but the terms here will have different values, as we will see below. Here the expansion is exact. The eigenvalue condition is $dD_2/d\mu_2 = 0$ and from (5.9.8) is

$$\sin[2k_0d\mu_{1m}] = 2\mu_{1m}/k_0dW \ . \quad (5.9.49)$$

If we define

$$x_m = 2k_0d\mu_{1m}; \quad \Delta = (k_0d)^{-2}W^{-1} \ , \quad (5.9.50)$$

then our eigenvalue equation can be written as

$$\sin x_m = \Delta x_m \ . \quad (5.9.51)$$

We find the x_m values by plotting $\sin x$ and Δx and determining their intersections. Since $\mu_1 = (W+\mu_2^2)^{1/2}$ and $\mu_2 : 0 \to \infty$ we have that $\mu_1 : W^{1/2} \to \infty$, so that $x = 2k_0d\mu_1$ runs from $2k_0dW^{1/2} \to \infty$. A sketch is represented in Fig. (5.12).

We illustrate the possible roots x_m, where "possible" means they must still be checked to satisfy the second derivative constraint in (5.9.45). For m even the roots are approximately periodically spaced, as they are for m odd. This differs from TSG The root positions are also clearly different from TSG The major difference, however, is that here there are only a *finite number of "possible" roots* since for some x, $\Delta x > 1$. In fact, to have any "possible" roots we must require that at $x = 2k_0 dW^{1/2}$ we must have $\Delta x < 1$ which means

$$k_0 d W^{1/2} > 2 \ . \tag{5.9.52}$$

For either low frequency, a narrow waveguide, or a small contrast with the bottom sound speed this condition may not be fulfilled. To see if all the x_m are roots, we have to check that the second derivative is positive. From (5.9.9) we use the fact that the first derivative vanishes and find that

$$\left.\frac{d^2 D_2}{d\mu_2^2}\right|_m = 2\frac{\mu_{2m}^2}{\mu_{1m}^2}\left(1 - \frac{\cos x_m}{\Delta}\right) \ . \tag{5.9.53}$$

For this to be positive for all values of Δ we require that $\cos x_m < 0$. From Fig.(5.12) we see that this can be satisfied for all x_m, m odd. These are thus our true virtual mode or resonance positions. The values of x_m, m even, do not satisfy this condition. The latter appear in other contexts also and are referred to as *antiresonances* [5.7].

To evaluate α_m we need $D_2(q_m)$. From (5.9.4) we have that

$$\begin{aligned} D_2(q_m) &= \mu_{1m}^2 \cos^2(x_m/2) + \mu_{2m}^2 \sin^2(x_m/2) \\ &= \mu_{1m}^2 - W \sin^2(x_m/2) \\ &= (W/4)(\Delta x_m^2 - 2 + 2\cos x_m) \ , \end{aligned} \tag{5.9.54}$$

where $\cos x_m < 0$. This yields

$$\alpha_m = \frac{4\,\mu_{2m}^2(\Delta - \cos x_m)}{\Delta \mu_{1m}^2 W(\Delta x_m^2 - 2 + 2\cos x_m)} \ . \tag{5.9.55}$$

The specific forms of the BLI's are the same as those in the TSG presentation, (5.9.32) for real virtual modes, and (5.9.44) for imaginary virtual modes. Here, however, the latter sum is finite and the terms are defined via our denominator minimization method, q_m and μ_{2m} using the values of x_m, m odd, from (5.9.51) (the true resonance positions), R_m using (5.9.18) but with $D_2(q_m)$ from (5.9.54), and α_m from (5.9.55). We do not repeat the specific functional forms here. We, of course, still have both the overall algebraic decay in range and the exponential decay of the modes.

Definition 3: A third approach to defining virtual modes can be found in Stickler and Ammicht ([5.9], [5.1]). We very briefly describe it here. They set as their eigenvalue equation

$$D_2(q_m) = 0 \ . \tag{5.9.56}$$

The poles thus occur in the complex plane, and considering D_2 as an even function of μ_2, we expand in a Taylor series in μ_2^2 as

$$D_2(q) \cong D_2(q_m) + \left[\mu_2^2 - \mu_{2m}^2\right] \left.\frac{dD_2}{d\mu_2^2}\right|_m \ . \tag{5.9.57}$$

This imples that g can be written as

$$g(z, z'; \mu_2^2) = \sum_m \frac{r_m}{\mu_2^2 - \mu_{2m}^2} \quad \mathrm{Im}\, \mu_{2m} > 0 \ , \tag{5.9.58}$$

where the residues can be written as

$$r_m = \sin(k_0 z' \mu_{1m}) \sin(k_0 z \mu_{1m}) \left\{ \left.\frac{dD_2}{d\mu_2^2}\right|_m \right\}^{-1} \ . \tag{5.9.59}$$

The resulting BLI can be written as

$$BLI \approx \frac{ik_0}{4\pi} \sum_m r_m M_m \ , \tag{5.9.60}$$

where the integrals are defined as

$$M_m = \int_{-\infty}^{\infty} H_0^{(1)}\left(k_0 r (K_2^2 - \mu_2^2)^{1/2}\right) \frac{\mu_2^2}{\mu_2^2 - \mu_{2m}^2}\, d\mu_2 \ . \tag{5.9.61}$$

A uniform asymptotic expansion of the integral is then carried out. It is uniform in the sense that it is valid as a mode passes through cutoff (i.e. as its position passes through a fixed branch point). The complex poles are located in the plane and the pole evaluation technique is compared to a numerical branch cut integration with good results. The fewer the poles or the shorter the range (really $k_0 r$) the worse the results. In this method both the leaky and shadow poles are important.

5.10 Modal Attenuation

In the previous sections we treated the analytic properties of the depth-dependent Green's function G_2 as a function of the dimensionless ray parameter q. The imaginary part of the eigenvalues of q contributed to the modal attenuation in range, as for example in the leaky mode sum in (5.7.20). Here we briefly consider the modal attenuation which can occur due to the flat boundaries at $z = 0$ and $z = d$. We can do this for a general depth-dependent profile $K(z)$ as well as including different densities in the waveguide and bottom. The eigenfunctions corresponding to (5.2.7) satisfy the equation

$$\frac{d^2}{dz^2} u_m + k_0^2 \left(K^2(z) - q_m^2\right) u_m = 0 \ , \tag{5.10.1}$$

where

$$q_m = \alpha_m + i\delta_m \ , \tag{5.10.2}$$

and δ_m is the modal attenuation coefficient. If we assume that in general the eigenfunctions and eigenvalues are complex then the complex conjugate of (5.10.1) is

$$\frac{d^2}{dz^2} u_m^* + k_0^2 \left(K^2(z) - q_m^{*2}\right) u_m^* = 0 \ . \tag{5.10.3}$$

If we include a density variability

$$\rho(z) = \begin{cases} \rho_1, & 0 \leq z \leq d \\ \rho_2, & z > d \end{cases} \ , \tag{5.10.4}$$

and form the quantity

$$\int_0^\infty \rho(z)[u_m^* \cdot (5.10.1) - u_n \cdot (5.10.3)] dz \ , \tag{5.10.5}$$

we get that

$$\int_0^\infty \rho(z) \frac{d}{dz} \left[u_m' u_m^* - u_m'^* u_m\right] dz = 4\, i k_0^2 \alpha_m \delta_m U_m \ , \tag{5.10.6}$$

where

$$U_m = \int_0^\infty \rho(z) |u_m|^2 dz \ . \tag{5.10.7}$$

Integrate the *lhs* of (5.10.6) by parts, use the definition

$$u_m = \begin{cases} u_{m1}, & 0 \le z \le d \\ u_{m2}, & z > d \end{cases} , \tag{5.10.8}$$

for the eigenfunctions in the two regions, and assume that both u_m and u'_m vanish at ∞. We can then write the attenuation coefficient as

$$\delta_m = \delta_m^0 + \delta_m^d , \tag{5.10.9}$$

where δ_m^0 is the attenuation coefficent at the $z = 0$ surface

$$\delta_m^0 = i\rho_1 \left(4k_0^2 \alpha_m U_m\right)^{-1} \left[u'_{m1} u^*_{m1} - u'^*_{m1} u_{m1}\right]\Big|_{z=0} , \tag{5.10.10}$$

and δ_m^d the bottom attenuation coefficient

$$\begin{aligned} \delta_m^d &= \left(4\, ik_0^2 \alpha_m U_m\right)^{-1} \left\{\rho_1 \left(u'_{m1} u^*_{m1} - u'^*_{m1} u_{m1}\right)\right. \\ &\quad \left. -\rho_2 \left(u'_{m2} u^*_{m2} - u'^*_{m2} u_{m2}\right)\right\}\Big|_{z=d} . \end{aligned} \tag{5.10.11}$$

We see that for Dirichlet or Neumann boundary conditions at the surface $z = 0$ there is no attenuation. Similarly for the usual continuity conditions at the bottom $z = d$ there is no attenuation.

Cases can arise where attenuation is present, however. An example is surface roughness. A second example was considered briefly in Chap. 3, that of a complex impedance boundary condition. For example, at the $z = 0$ surface if instead of Dirichlet or Neumann boundary conditions we have

$$u_{m1} = \gamma u'_{m1} ; \quad u^*_{m1} = \gamma^* u'^*_{m1} , \tag{5.10.12}$$

which is a complex impedance boundary condition, and if we define

$$\gamma = iZ/\omega\rho_1 , \tag{5.10.13}$$

where Z is the boundary impedance, it follows from (5.10.10) that

$$\begin{aligned} \delta_m^0 &= i\rho_1 \left(4\, k_0^2 \alpha_n U_n\right)^{-1} 2i\, \mathrm{Im}\delta |u'_{n1}|^2 \\ &= -Z|u'_{m1}|^2 \left(2k_0^2 \alpha_m U_m \omega\right)^{-1} , \end{aligned} \tag{5.10.14}$$

which contributes variable absorption depending on frequency and density. The absorption is also of course different for each mode.

5. Appendix

5.A Proof of the Contour Representation

In this appendix we prove the contour representation (5.2.2). We begin by noting that the problem is separable and assume a solution of the form

$$G(\boldsymbol{r}, \boldsymbol{r}') = \alpha \int_{\Gamma_\lambda} G_1(r;\lambda) G_2(z, z';\lambda) d\lambda \ , \qquad (5.A.1)$$

where we must find the constant α and the contour Γ_λ. Differentiate (5.A.1) to yield

$$\begin{aligned} \left[\nabla^2 + k_0^2 K^2(z)\right] G &= \alpha \int_{\Gamma_\lambda} \Big\{ G_2(z, z';\lambda) \frac{1}{r}\frac{d}{dr}\left(r\frac{dG_1}{dr}\right) \\ &\quad + G_1(r;\lambda)\left[\frac{d^2}{dz^2} + k_0^2 K^2(z)\right] G_2 \Big\} d\lambda \ . \qquad (5.A.2) \end{aligned}$$

We assume (5.2.3) and (5.2.4) so that (5.A.2) becomes

$$\begin{aligned} \left[\nabla^2 + k_0^2 K^2(z)\right] G &= \alpha \Big\{ -\frac{\delta(r)}{2\pi r} \int_{\Gamma_\lambda} G_2(z, z';\lambda) d\lambda \\ &\quad - \delta(z - z') \int_{\Gamma_\lambda} G_1(r;\lambda) d\lambda \Big\} \ . \qquad (5.A.3) \end{aligned}$$

We have the following completeness relationships (which are proved below)

$$\delta(z - z') = \frac{k_0^2}{2\,\pi i} \int_{\Gamma_2} G_2(z, z';\lambda) d\lambda \ , \qquad (5.A.4)$$

where Γ_2 encloses all singularities of G_2 on the proper sheet of the complex λ-plane in a counter-clockwise direction, and

$$\frac{\delta(r)}{2\pi r} = \frac{k_0^2}{2\pi i} \int_{\Gamma_1} G_1(r;\lambda) d\lambda \ , \qquad (5.A.5)$$

where Γ_1 encloses all singularities of G_1 on the proper sheet of the complex λ-plane in a clockwise direction. Substituting (5.A.4) and (5.A.5) into the *rhs* of (5.A.3) we get

$$\left[\nabla^2 + k_0^2 K^2(z)\right] G = -\alpha \frac{2\pi i}{k_0^2} \frac{\delta(r)\delta(z-z')}{2\pi r} \ , \tag{5.A.6}$$

which satisfies (5.2.1) if we choose $\alpha = k_0^2/2\pi i$.

We first prove the completeness relation (5.A.4). The differential equation for G_2 is

$$\frac{d^2}{dz^2} G_2 + k_0^2(K^2(z) - \lambda) G_2 = -\delta(z-z') \ . \tag{5.A.7}$$

If $K(z) = K_1$ or K_2 a constant the proof is simple. Consider the slightly more general equation

$$\frac{d^2}{dz^2}\psi + k_0^2(K^2 - \lambda)\psi = f(z) \ . \tag{5.A.8}$$

If $f(z) = -\delta(z-z')$ then for whatever choice of K, ψ is G_2. Expand ψ and f in orthonormal eigenfunctions u_n satisfying the equation (λ_n are eigenvalues)

$$\frac{d^2}{dz^2} u_n + k_0^2(K^2 - \lambda_n) u_n = 0 \ . \tag{5.A.9}$$

These expansions are

$$\psi(z) = \sum_n \psi_n(\lambda_n) u_n(z, \lambda_n) \ , \tag{5.A.10}$$

and

$$f(z) = \sum_n f_n(\lambda_n) u_n(z, \lambda_n) \ . \tag{5.A.11}$$

For simplicity, we assume a discrete spectrum and generalize to a continuous spectrum below. Using the orthonormality of the eigenfunctions we solve (5.A.10) and (5.A.11) to get

$$\psi_n(\lambda_n) = \int \psi(z) u_n^*(z, \lambda_n) dz \ , \tag{5.A.12}$$

and

$$f_n(\lambda_n) = \int f(z) u_n^*(z, \lambda_n) dz \ . \tag{5.A.13}$$

(Note that if we have a continuous spectrum, the orthonormality of the eigenfunctions must be proved.) Next substitute (5.A.10) and (5.A.11) into (5.A.8) to get

$$\sum_n \psi_n \left\{ u_n'' + k_0^2(K^2 - \lambda)u_n \right\} = \sum_n f_n u_n \ . \tag{5.A.14}$$

Substitute for the second derivative from (5.A.9) and again use orthonormality to show that

$$\psi_n = f_n / k_0^2(\lambda_n - \lambda) \ . \tag{5.A.15}$$

Then (5.A.10) becomes

$$\begin{aligned} \psi(z) &= \sum_n \frac{u_n(z,\lambda_n)}{k_0^2(\lambda_n - \lambda)} f_n \\ &= \sum_n \frac{u_n(z,\lambda_n)}{k_0^2(\lambda_n - \lambda)} \int f(z'') u_n^*(z'',\lambda_n) dz'' \ , \end{aligned} \tag{5.A.16}$$

where the latter follows from (5.A.13). If also $f(z'') = -\delta(z'' - z')$ then $\psi = G_2$ and (5.A.16) becomes

$$G_2(z,z';\lambda) = \sum_n \frac{u_n(z,\lambda_n) u_n^*(z',\lambda_n)}{k_0^2(\lambda - \lambda_n)} \ , \tag{5.A.17}$$

and if we integrate (5.A.17) in a counter-clockwise manner we get

$$\frac{1}{2\pi i} \int_{\Gamma_2} G_2(z,z';\lambda) d\lambda = \frac{1}{k_0^2} \sum_n u_n(z,\lambda_n) u_n^*(z',\lambda_n) \ , \tag{5.A.18}$$

where Γ_2 surrounds all the singularities of G_2. Further we have from (5.A.11) that

$$-\delta(z - z') = \sum_n f_n(\lambda_n) u_n(z,\lambda_n) \ , \tag{5.A.19}$$

and from (5.A.13)

$$f_n(\lambda_n) = -u_n^*(z',\lambda_n) \ . \tag{5.A.20}$$

Combining (5.A.19) and (5.A.20) we can write (5.A.18) as

$$\delta(z - z') = \frac{k_0^2}{2\pi i} \int_{\Gamma_2} G_2(z,z';\lambda) d\lambda \ , \tag{5.A.21}$$

which is the completeness relation (5.A.4).

To prove (5.A.5) we note that the only essential difference in (5.2.3) and (5.2.4) insofar as the completeness proof is concerned is the sign of λ. We can thus replace

$$\delta(z-z') \rightarrow \delta(r)/2\pi r$$

$$\lambda \rightarrow -\lambda$$

$$G_2 \rightarrow G_1$$

in (5.A.21) and immediately derive the result

$$\frac{\delta(r)}{2\pi r} = \frac{k_0^2}{2\pi i}\int_{\Gamma_1} G_1(r;\lambda)d\lambda \ , \tag{5.A.22}$$

where Γ_1 encircles all the singularities of G_1 now in a clockwise manner. This is just (5.A.5) and the proof is complete.

References

Chapter 1

1.1 G. Birkhoff, G-C. Rota: *Ordinary Differential Equations* (Ginn and Company, Boston 1962)

1.2 E.A. Coddington, N. Levinson: *Theory of Ordinary Differential Equations* (McGraw-Hill, New York 1955)

1.3 R. Courant, D. Hilbert: *Methods of Mathematical Physics* (Interscience, New York 1953) Vol. I

1.4 J.A. DeSanto: Derivation of the acoustic wave equation in the presence of gravitational and rotational effects. J. Acoust. Soc. Am. **66**, 827-830 (1979)

1.5 R. Haberman: *Elementary Applied Partial Differential Equations* (Prentice-Hall, Englewood Cliffs, NJ 1983)

1.6 L.D. Landau, E.M. Lifshitz: *Fluid Mechanics* (Addison-Wesley, Reading, Massachusetts 1959)

1.7 P.M. Morse, K.U. Ingard: *Theoretical Acoustics* (Princeton Univ. Press, Princeton, NJ 1986)

1.8 J.W. Strutt (Baron Rayleigh): *The Theory of Sound* (Dover, New York 1945) Vols. 1 and 2

1.9 I. Tolstoy: *Wave Propagation* (McGraw-Hill, New York 1973)

1.10 E.T. Whittaker, G.N. Watson: *Modern Analysis* (Cambridge Univ. Press, Cambridge 1962)

Chapter 2

2.1 R. Courant, D. Hilbert: *Methods of Mathematical Physics* (Interscience, New York 1953) Vol. 1

2.2 A.J. Devaney, G.C. Sherman: Plane-Wave representations for scalar wave fields. SIAM Review **15**, No. 4, 765-786 (1973).

2.3 B. Friedman: *Principles and Techniques of Applied Mathematics* (Wiley, New York 1962)

2.4 P.M. Morse, H. Feshbach: *Methods of Theoretical Physics* (McGraw-Hill, New York 1953) Parts I and II

2.5 I. Stakgold: *Green's Functions and Boundary Value Problems* (Wiley, New York 1979)

2.6 H.W. Wyld: *Mathematical Methods for Physics* (Benjamin, Reading, MA 1979)

Chapter 3

3.1 M. Born, E. Wolf: *Principles of Optics,* 6th edn. (Pergamon, New York 1980)

3.2 L.M. Brekhovskikh: *Waves in Layered Media* (Academic, New York 1960)

3.3 L.M. Brekhovskikh, O.A. Godin: *Acoustics of Layered Media I* Springer Ser. Wave Phen., Vol. 5 (Springer, Berlin, Heidelberg 1990)

3.4 J.A. DeSanto, G.S. Brown: Analytical techniques for multiple scattering from rough surfaces. *Progress in Optics* **23**, 1-62 (North-Holland, Amsterdam 1986)

3.5 D.S. Jones: *Acoustic and Electromagnetic Waves* (Clarendon, Oxford 1986)

Chapter 4

4.1 K. Aki, P.G. Richards: *Quantitative Seismology, Theory and Methods* (Freeman, New York 1980) Vol. I

4.2 N. Bleistein, R.A. Handelsman: *Asymptotic Expansions of Integrals* (Dover, New York 1986)

4.3 L.M. Brekhovskikh: *Waves in Layered Media* (Academic, New York 1960)

4.4 A. Erdelyi: *Asymptotic Expansions* (Dover, New York 1956).

4.5 L.B. Felsen, N. Marcuvitz: *Radiation and Scattering of Waves* (Prentice Hall, Englewood Cliffs, NJ 1973)

4.6 I.S. Gradshteyn, I.M. Ryzhik: *Tables of Integrals, Series, and Products* (Academic, New York 1965)

4.7 W. Magnus, F. Oberhettinger, R.P. Soni: *Formulas and Theorems for the Special Functions of Mathematical Physics* (Springer, New York 1966)

4.8 E.T. Whittaker, G.N. Watson: *Modern Analysis* (Cambridge Univ. Press, Cambridge 1962)

Chapter 5

5.1 E. Ammicht, D.C. Stickler: Uniform asymptotic evaluation of the continuous spectrum contribution for a stratified ocean. J. Acoust. Soc. Am. **76**, 186-191 (1984)

5.2 G. Birkhoff, G-C. Rota: *Ordinary Differential Equations* (Ginn and Company, Boston 1962)

5.3 L. Brekhovskikh, Yu. Lysanov: *Fundamentals of Ocean Acoustics* 2nd edn., Springer Ser. Wave Phen., Vol. 8 (Springer, Berlin, Heidelberg 1991)

5.4 J.A. DeSanto: Theoretical Methods in Ocean Acoustics. *Ocean Acoustics*, ed. by J.A. DeSanto, Topics Curr. Phys., Vol. 8 (Springer, Berlin, Heidelberg 1979) pp. 7-77

5.5 F.R. DiNapoli, R.L. Deavenport: Numerical models of underwater acoustic propagation. *Ocean Acoustics*, ed. by J.A. DeSanto, Topics Curr. Phys., Vol. 8 (Springer, Berlin, Heidelberg 1979) pp. 79-157.

5.6 W.M. Ewing, W.S. Jardetzky, F. Press: *Elastic Waves in Layered Media* (McGraw-Hill, New York 1957)

5.7 R.G. Newton: *Scattering Theory of Waves and Particles* (McGraw-Hill, New York 1966)

5.8 C.L. Pekeris: Theory of propagation of explosive sound in shallow water. Geol. Soc. Am. Mem. **27** (1948)

5.9 D.C. Stickler, E. Ammicht: Uniform asymptotic evaluation of the continuous spectrum for the Pekeris model. J. Acoust. Soc. Am. **67**, 2018-2024 (1980)

5.10 C.T. Tindle, A.P. Stamp, K.M. Guthrie: Virtual modes and the surface boundary condition in underwater acoustics. J. Sound Vib. **49**, 231-240 (1976).

5.11 A.O. Williams, Jr.: Discrete, continuous and virtual modes in underwater sound propagation. Tech. Rept. ARL-TR-76-40, Applied Research Laboratory, Austin, TX (1976)

Index